composites
A DESIGN GUIDE

composites

A DESIGN GUIDE

Terry Richardson

Industrial Technology Department
Northern State College
Aberdeen, South Dakota

INDUSTRIAL PRESS INC.

Library of Congress Cataloging-in-Publication Data

Richardson, Terry L.
    Composites: a design guide.
    Bibliography: p.
    Includes index.
    1. Polymeric composites. I. Title.
TA418.9.C6R53 1987   668.9   87-609
ISBN 0-8311-1173-9

Industrial Press Inc.
200 Madison Avenue
New York, New York 10016

FIRST PRINTING

Composites: A Design Guide

Composition by Crane Typesetting, Barnstable, MA. Printed and bound by Quinn Woodbine, Woodbine, N.J.

2 4 6 8 9 7 5 3

Preface

Composites: A Design Guide is an introductory, somewhat simplified, basic textbook appropriate for a first course in polymer composites. It is designed for those who will need to select the correct material process for composite design requirements.

Charles Kettering used to say "Chance favors the prepared mind." This book provides the latest information and breakthroughs on materials, processes, and applications of polymeric composites. If Alvin Toffler is correct, we must know how to design, fabricate, and use composite materials in a third-wave society. New, innovative, and creative ideas are yet to be realized with composite materials.

Despite their existence for a number of years, composites have occupied an ambiguous and ill-defined position as an important class of engineering materials. Few other materials are expected to experience as rapid a growth as composites. Metallic and ceramic composite materials are also an important engineering material, but discussion is limited to polymeric matrix composites.

The plastics and composite industry generally has experienced a shortage of personnel with knowledge about composite manufacturing design and productivity. This lack of knowledge has sometimes led to a combination of overdesign and product failures. Therefore, this text should be of interest to all those engaged in learning, training, manufacturing, or designing with polymeric composites and should be an important reference for all those who will need to select the correct material and process for selected design requirements.

The confidential and proprietary details outlined or other information provided is intended only as a guide. It is not to be taken as a license to operate or as a recommendation to infringe on any patents.

Contents

viii

Chapter 1. What Are Composites?

Introduction

The development of engineering through the centuries was dependent on the discovery and availability of new structural materials. As new materials were used, their limitations became evident and new natural or manufactured resources were tried.

Wood was light and tough but would burn. At one time wood was the most widely used engineering material because of a number of outstanding properties. However, as a product of nature, it has one serious property flaw. Modern designs and specifications demand that materials be reliable from batch to batch. While we have come to expect this reliability of properties in metals, ceramics, and plastics, wood properties vary from batch to batch because it is not an isotropic (same properties in all directions) substance but a complex material. Wood is highly anisotropic with its greatest strength and stiffness parallel to the grain structure. Early materials such as metals and ceramics would be considered isotropic. The addition of ribs, reinforcements, laminates, or corrugations will make these materials artificially anisotropic in strength.

Ceramics and some metals were strong but heavy. The technology of ceramic ware and glass making has remained fundamentally the same for centuries. Products are essentially the same as those used by the Chinese thousands of years ago. Modern engineering and the process of ceramics and glasses are providing a new generation of high-performance materials from the shielding on spacecraft to automobile engine components.

Naturally occurring polymers such as shellac, resin, asphalt, tar, and lignin were easily shaped but have little strength.

Society has always wanted and continues to seek materials that are strong, tough, and light. In this quest, it was discovered that copper could be made harder by mechanically shaping the homogeneous material. Gold was highly prized because it was found occurring in nuggets and could easily be shaped into useful products. The addition of other metals or combinations of metals improved the durability and strength of both copper and gold. The concept of alloying metals to improve or satisfy the increasing demands for superior materials was learned centuries ago. Even today, there have been improvements in alloying steel, aluminum, titanium, and other metals. You might say that product development has passed through several periods of significant improvement in naturally occurring materials: the bronze age, the steel age, and the plastics age, which occurred during the middle of this century. Although the concept of combining two materials together with complementary properties was known early in human history, modern designers, engineers, and others were slow to capitalize on the merits of this concept. In fact, it might be said that the plastics industry was in the bronze age of plastics development during the 1970s. Up until that time, most of the development was directed toward discovery of new monomers or homopolymers. Work began by the chemists in 1830 produced the first synthetic plastics. These materials were developed to provide improvements over natural polymers. Some were discovered by chance, others remained laboratory curiosities or undeveloped for years.

Modifying natural sources of resins was a natural step toward improvement of polymeric materials. Modifications in cellulose, protein, gelatin, and latex were important contributions to the development of new polymers and product applications. In 1948, acrylonitrile–butadiene–styrene (ABS) plastics were produced. The proportions of the three ingredients may vary, which accounts for the great number of possible properties. Some classify ABS as a family of plastics such as polystyrene, polyvinyl chloride, and others. They are, however, terpolymers (*ter* meaning three) of three monomers. Today, polymer chemists and engineers are combining (or in a sense alloying) many synthetic polymers to attain improved properties or processing at lower cost.

Styrene monomers are added to acrylic to lower costs. Many barrier plastics are a combination of acrylonitrile–butadiene–styrene. These materials stop or limit the passage of gases through the container wall. There is little doubt that we will continue to see improved materials with continued copolymerization and terpolymerization of polymeric materials.

The idea of combining two or more different materials resulting in a new material with improved properties is much older than modifying the polymer chain. Humans discovered long ago that **composite** materials should have the combined advantages with superior performance in comparison with each individual material. For more than 3000 years, humans have been putting fibers into a binder or matrix body. For example, in biblical times, the Pharoahs of Egypt and the ancient Incan and Mayan societies knew that plant fibers helped strengthen and prevent bricks and pottery from cracking, and moss is used by the Eskimos to strengthen ice. Even Leonardo da Vinci conceived and used composite parts. He understood the advantages of laminating wood, metals, paper, and fabric to produce a material with new characteristics and properties.

Ancient sarcophagi were laminated with cross-banded and cross-plied wood veneers to form beautiful shapes that would withstand centuries. The merits of plywood and laminated beams are well known. In Figure 1-1, laminated wooden beams support the twisted steel beams in the aftermath of fire. Cross-plies help overcome the dimensional instability and the variations of wood. Early technicians knew the concepts of producing composite materials that were artificially anisotropic.

Medieval armorers realized that a superior weapon could be made by producing a composite bow of wood and sinew (filaments of bull tendon) all filament wound (overlapped) with silk fiber. Early cannons were reinforced by wrapping a high-tension wire around the cast barrel to withstand greater internal exploding pressures. Some of the early firearms and swords were produced by putting several different steels together in laminated layers. For example, the famous Samurai sword was produced by layering different steels and forging them together. Because of the repeated folding and reshaping of the multilayered metal composite, these fine swords are estimated to contain as many as 8,000,000 layers of steel!

Today, a mixture of sand, gravel, and cement is used to form

Figure 1-1. Composite (wood laminated) beams support twisted steel beams in aftermath of fire. (Courtesy: American Institute of Timber Construction)

concrete. The strength of steel reinforcing bars and the rigidity of the concrete matrix are widely used today to make superior structural members.

A final example of composites is the wheel, which has evolved slowly from wood to wire to steel to aluminum. Composite fibrous glass wheels are 20% lighter than aluminum.

Many people continue to treat metals, ceramics, woods, and polymers as separate materials with relatively little in common. Some school curriculums still have classes in woods, metals, plastics, and ceramics. Although there may be some justification because of laboratory requirements, a systems concept to material usage is justified. There are increased demands for new materials, products, and processing techniques. Advances in the science and technology of materials have provided us with a new potential for the future. Many of the demands made on modern materials are so severe that individual materials no longer perform or have the desired properties, so it is frequently necessary to combine several materials into a composite. As you have seen, a composite is formed when two or more materials are combined, usually with the intent of achieving better results than can be obtained with a single, homogeneous material. We must learn to process, design, and use these new materials.

The processing techniques, designs, and material science of uniform, homogeneous substances are complicated when a mixture of materials is considered. We cannot rely solely on the technology of homogeneous materials science, design, or processing. Using composite materials may require radical departures from establised processing and design techniques. Technicians, engineers, chemists, metallurgists, ceramists, designers, processors, manufacturers, and others must learn to cope with these new materials called **composites**.

Definition of Composites

The word **composite** implies that it is a material composed of two or more distinct components. These are among the oldest and newest materials. Naturally occurring composites are bone, bamboo, feathers, natural fibers, and wood. Bone is a composite of protein (collagen) and mineral (apatite). Bamboo is a cellulose reinforced by silica. This combination makes bamboo a hard material with high impact strength. The cellulose cell structure of wood and fiber is bound together with lignin, a natural polymeric substance. Wood cells are natural composite structures within themselves.

The word **composite** has evolved over a period of years. Some consider it a system or process of combining two or more reinforcing materials in a matrix binder. Others consider it to be a new material having characteristics derived from its processing and from its microstructure.

A precise definition of composite must be divided into two basic forms: (1) composite materials and (2) composite structures. **Composite materials** are composed of a reinforcing structure, surrounded by a continuous matrix. The structure must be capable of arbitrary variation to be considered a composite material. **Composite structures** exhibit a discontinuous matrix, in which the dissimilar materials are not capable of arbitrary variation. Although wood veneers and a matrix of adhesive may act synergistically to improve physical properties, they must be classified as composite structures. Most laminates, including plywood, safety glass, metal clads, and laminated metals, fall into this form. In much of the literature, the term **structural composites** is used to describe the application of the composite material, such as high-strength, reinforced plastics bulkheads. The actual part may be a **composite material**.

In both forms, components (reinforcements and matrix) do not normally dissolve cohesively together but do contribute to a synergistic property change. It is beyond the scope of this text to discuss all forms of composites and those that do not conform to a standard definition.

Classification of Composites

There are three major classifications of composites: (1) fibrous, (2) laminar, and (3) particulate.

Fibrous Composites

Fibrous composites are composed of reinforced fibers in a matrix. It is this classification into which reinforced plastics are placed. Many refer to reinforced plastics and elastomers as composite materials. (See Reinforced Plastics.)

Fibers are small in diameter and when pushed axially will bend easily. Although they may have outstanding tensile strength, they must be supported to keep individual fibers from bending and buckling.

Metal, glass, and ceramic fibers placed in a metallic matrix offer higher temperature capability, strength, and stiffness than is possible using a plastic matrix.

During the 1970s, many in the reinforced plastics industry began to talk about "high-strength composites" and "advanced composites." These terms have been generally associated with reinforcements that offered stiffer, stronger composites than the traditional composite of glass fibers in a polyester matrix. During this period, there was a rapid growth or use of composite materials. Associated with these new composites were exotic new reinforcements of carbon, boron, graphite, tungsten, and others. It is true that these materials did offer advantages of high stiffness, low density, good fatigue resistance, and excellent thermal and dimensional stability over the more traditional glass composites of the day. Perhaps this usage will disappear in time, since it assumes that these materials are "advanced" over other composites. The same argument has been made about the term "engineered plastics" or "engineered composites." We must assume that others are not engineered!

Laminar Composites

Laminar composites are composed of layers of materials held together by the matrix binder. Sandwich and honeycomb components as well as the term *high-pressure* laminates are included. Wooden laminates, plywood, and some combinations of metal foils, glasses, plastics, films, and paper are laminar composites. Some ceramic and metallic composites also fall into this classification.

Particulate Composites

Particulate composites consist of particles dispersed in a matrix. These particles are sometimes divided into two subclasses: (1) skeletal, which consists of a continuous skeletal structure filled with one or more additional materials; and (2) flake, which consists generally of flat flakes oriented parallel to each other. Particles may have any shape, configuration, or size. These particulates may be powdered, beads, rods, crystalline, amorphous, or whiskered. They may be metallic, ceramic, manmade, or natural materials. The distinction between fillers and reinforcing agents will be discussed later. Concrete and wood particleboard are two familiar examples of particulate composites. Metallic flakes have been added to improve electrical properties and provide some degree of radiation shielding in polymer composites. Pieces of ceramic particles are placed in a metallic matrix and used as tough, abrasion resistant cutting tools.

MATRICES

The matrix is the material that gives body and grips or holds the reinforcements of the composite together, and is usually of lower strength than the reinforcement. The matrix must be capable of being forced around the reinforcement during some stage in the manufacture of the composite. There are a number of matrix materials available, including carbon, ceramic, glass, metal, and polymer. This text will be devoted primarily to polymer matrices.

Carbon Matrix

Two familiar forms of carbon are diamond and graphite. There are many rich sources of carbonaceous material such as coal, petroleum, and coke. Because carbon and graphite have a high heat capacity per unit weight (mass), they are selected as ablative materials. As a carbon matrix, it is used as rocket nozzles, ablative shields for reentry vehicles, and clutch and brake pads in aircraft. Many of the carbon matrix composites are produced by pressing powders or slurry castings of carbon. The carbon is then heated to above 2500° C to transform it to a graphite form. Carbon whiskers or fibers are generally used in the matrix, depending on the processing technique and desired properties.

Ceramic Matrix

Ceramic materials are crystalline and have properties similar to glass. The matrix is brittle even with attempts at annealing. Sintered and cast pieces shrink during firing. Carbon, ceramic, metal, and glass fibers are used in ceramic matrices. Many of the ceramic composites experience degradation of the fiber–matrix bonds as a result of oxidation and further stress cracking. Ceramic composites are used in rocket engine parts and protective shields.

Glass Matrix

Glass and glass–ceramic composites are characterized with an elastic modulus generally lower than the reinforcement. Crack propagation in the matrix may cause deterioration of the fibers. Carbon and metal oxide fibers are the most common reinforcements. Strength at high service temperatures is the most unique attribute of glass and ceramic matrices. Heat resistant parts for engines, exhausts, and electrical components are their primary applications.

Metal Matrix

For demanding high-temperature usage in oxidizing environments, metal matrices of iron, nickel, tungsten, titanium, aluminum, and magnesium are used in place of polymers. The low-density metals have long been a favorable choice in aircraft and aerospace designs. According to one source, metal composites may be classified according to their degree of reaction between the reinforcement and the matrix phases. In Class I, the reinforcement and matrix are insoluble; there is little chance that degradation will affect service life of the part. Class II components exhibit some solubility; over a period of time and during processing, the interaction will alter the physical properties of the composite. The most critical situation, and one representing technical processing problems, is Class III composites. (See Table 1-1.)

Aluminum and magnesium matrices are most commonly used with boron, carbon, graphite, and aluminum oxide fiber reinforcements. These are favored because of their strong anisotropy of mechanical and physical properties. The long history of fabrication technology and availability of aluminum make it most popular. Most fibers must be coated or treated in order to promote the best adhesion. Fabrication methods must consider interaction between components and involve alignment and spacing of reinforcements for optimum properties and service. These composites find applications in compression panels and stiffeners in missiles and other craft. Silica fibers, silicon carbide whiskers, and various metal wires have been used to reinforce aluminum.

Titanium is the heaviest of the light metals and also one of the strongest. It has excellent corrosion resistance and can withstand temperatures of more than 1000° C. The most important reinforcements are boron, silicon carbide, and silicon-carbide-coated boron.

Alloy matrices of tungsten, nickel, cobalt, iron, and other formulations are often used for high-strength, stiff, oxidation-resistant, high-temperature applications. Because of the processing techniques and extreme-temperature applications, metal-fiber reinforcements of molybdenum, tungsten, tantalum, and niobium have been used.

Table 1-1. Classification of Metal Matrix Composite Systems[a]

Class I	Class II	Class III
Copper–tungsten	Copper (chromium)–	Copper (titanium)–
Copper–alumina	tungsten	tungsten
Aluminum–BN-coated B	Columbium–tung-	Aluminum–carbon
Magnesium–boron	sten	Titanium–alumina
Aluminum–boron	Nickel–carbon	Titanium–boron
	Nickel–tungsten	Titanium–silicon
		carbide
		Aluminum–silica

[a]Modified from Metcalfe.

Polymer Matrix

Polymers were a logical choice for matrices. They were found occurring in nature as amber, pitch, and resin. Some of the earliest composites were layers of fiber, cloth, and pitch. Elastomers of natural rubber including gutta percha were reinforced with fiber and fabrics before 1850. At the turn of the century, synthetic polymers became important. Compounds of urea-formaldehyde, asbestos, and other formulations became commonly used in the industry as phenolic laminates. By the end of World War II, a variety of reinforcing agents and resins were used successfully in structural applications for boats, cars, appliance housing, trays, and aircraft. Some of the first applications were for replacement of wood in fuselage design. The famous wooden fighters and gliders of this period encountered many structural problems; composite structures were used to meet the increasing structural requirement of aircraft designs. Sandwich, honeycomb core, and laminates were used. See Figure 1-2.

Figure 1-2. Composite car; 1953 Chevrolet Corvette was first composite production body.

Polymers were selected because they were easily processed and offered good mechanical and dielectric properties. Most wet the reinforcements well, resulting in good adhesion. Although polymers have lower softening points than metals, they are low-density materials. It is because of the relatively low processing temperatures and production techniques that many organic reinforcements may be used. Natural and synthetic fibers including silk, cotton, wool, cellulosics, polyester, polyamide, acrylic, and olefins are used. Various inorganic, nonmetallic fibers of glass, carbon, asbestos, graphite, carbides, and oxides are in service. Glass fibers and other reinforcement forms are widely used in industrial and automotive applications. Metallic fibers of steel, copper, aluminum, titanium, tungsten, and nickel find applications in numerous products. Steel-belted radial tires with an elastomeric matrix are important composite materials.

REINFORCED VERSUS REINFORCING

Composites are often referred to as reinforced plastics (RP), fiberglass reinforced plastics (FRP), and filled molding compounds. The term composite is sometimes used to describe any laminated material. The literature may indicate that when the reinforcing material is placed in layers or plies, the resulting product is called a reinforced plastic laminate.

All discussion in this text will consider the term **reinforced** to mean a system or form of a composite. It will be used to indicate that a special reinforcing agent has been added to a polymer matrix. It does not refer to a specific polymer or processing technique. The word **reinforcing** will be used to describe a process or the verb tense of strengthening and/or making stronger.

All fibrous and particulate composites classes will be placed into the processing category of reinforcing. (See Reinforcing Processes.) Reinforced plastics are composite materials with a polymeric matrix and a reinforcing agent, either alone or together, of fibers or particles.

LAMINATED VERSUS LAMINATING

A similar problem of confusion exists in the use of the term **laminated**. In this text, the term laminated will be used to describe a plastics form or product. **Laminating** will be used to describe a process. All laminar composite classes of materials will be placed into the processing category of laminating. (See Laminating Processes and Figure 1-3.)

Figure 1-3. Familiar composite laminates used in transportation, communications, electrical circuit boards, and appliance industries. (Spaulding Fibre Co.)

Chapter 2.
Which Polymers to Select?

Introduction

By now we should have a general notion of what a composite is and how it differs from homogeneous or monolithic materials. The polymer matrix can be thought of as the "body" of the composite: It gives the composite its bulk form. This does not imply that it is the major component. Some composites may be composed of less than 10% matrix.

Laminar composites of sandwich, honeycomb, foam materials, foil, or other layers (lamina) give the composite its form. Only a small amount of matrix may be required to bond or hold the layers together.

The relative role of the matrix is to help transfer stress to the reinforcing agents, lower cost, or improve physical appearance or environmental properties. The other constituents, sometimes called the structural constituents, determine the mechanical and physical properties of the composite's internal structure.

Many considerations must be carefully viewed in selecting the proper plastics matrix. Design criteria are usually one of the first. The mechanical, thermal, and chemical properties of the matrix may limit some designs, while processing techniques and the matrix cost may limit others. If the polymer does not adhere well to the reinforcements or there are voids in the constituents, this will have a deleterious effect on shear strength, flexural strength, and other properties.

It should be evident that selection of the composite's matrix will be one of the constituents that determine the composite's performance.

Polymers are characterized as long chains of repeating molecules. The word **polymer** is derived from the Greek **poly**, meaning many, and **meros**, meaning unit or part. They are also called resins, macromolecules (from the Greek **macros**, meaning long), elastomers, and plastics. In this text, the word **polymer** will be used in a general sense to encompass all large molecular hydrocarbon materials. Most will refer to those repeating molecules based on the element carbon.

The term **polymer** actually refers to any material with a long chainlike structure of repeating molecules having covalent bonds. **Resins** are the organic substances from which paints, varnishes, plastics, and elastomers are produced. A resin is not a plastic unless and until it has become a "solid in the finished state." The term has been (incorrectly) used to refer to any material whose molecules are polymers. The term **macromolecule** may be misleading because it does not convey the concept of a repeating unit. There are many chemicals that are composed of long molecules. **Elastomer** is used to describe any polymer that is capable of being stretched more than 200%. The ASTM definition of an elastomer is "a polymeric material which at room temperature can be stretched to at least twice its original length and upon immediate release of the stress will return quickly to approximately its original length." The word **plastic** is an adjective, meaning pliable and capable of being shaped by pressure. Plastic is often incorrectly used as the generic word for the plastics industry and its products. The confusion arose because the word may be used as a noun or an adjective with both singular and plural meanings. The word **plastics** comes from the Greek word **plastikos**, meaning "to form or fit for molding." The Society of the Plastics Industry has defined plastics as "any one of a large and varied group of materials consisting wholly or in part of combinations of carbon with oxygen, nitrogen, hydrogen and other organic or inorganic elements which, while solid in the finished state, at some stage in its manufacture is made liquid, and thus capable of being formed into various shapes, most usually through the application, either singly or together, of heat and pressure."

Polymer Classifications

There are patents for more than 30,000 polymers or polymer systems. This large number of natural and synthetic materials have been classified in several ways: (1) source, (2) heat reaction, (3) polymerization reactions, (4) molecular structure, and (5) crystal structure.

Relative to Source

There are three principal sources of polymers. Polymers from natural sources are asphalt, amber, resin, shellac, lignin, and others. Composites of asphalt, tar, and pitch have been used for roofing materials for more than 200 years. Wood and wood cells are natural composite structures, with the matrix of wood being composed largely of lignin. Lignin and other polymer matrices are used to produce pressed hardboard and particleboard.

Cellulose and protein-derived polymers come from modified natural sources. Some prefer to class them as chemical-derived polymers from natural sources. Cellulose from cotton and other plant wastes gave Alexander Parks of England and John Hyatt of the United States important sources for the production of cellulose-based plastics. Today, more than ten basic resins represent the cellulose family of polymers. Protein is another important source of polymeric materials. Milk, soybeans, peanuts, bones, and other wastes are protein sources.

Casein (produced from milk) was used as early as 1890 to produce laminated wooden beams and arches. Gelatin, a protein derived from animals, is used as a matrix in some candies, cosmetics, meats, and medicines. The polymer rubber and similar materials such as gutta percha have been used for numerous elastomer products. Fibrous reinforcements of cellulose (Rayon) and elastomer are composites still used today. Millions of tires made from these reinforced materials have been sold.

There was increased interest in the development of natural and modified natural sources as concern grew about petroleum supplies during the 1970s and 1980s. Although there is still a market for these polymer sources, they have largely been replaced by synthetic sources.

The three major sources of chemicals for synthetic plastics are petroleum, coal, and agriculture. Hydrocarbons from petroleum have been the largest, most important source of polymers for more than 50 years. During the distillation process, petroleum yields thick asphalt, kerosine, diesel fuel, gasoline, and benzene liquids. Methane, ethylene, propane, and propylene gases are also derivatives. These light hydrocarbon gases are important building blocks of many polymers. Coal is a rich source of most of these same building blocks and because of the limited supplies of petroleum in Europe, coal and coal tars have historically been major sources of feedstocks for polymer production.

As we continue to deplete the world's reserves of petroleum and coal, we must seek other sources to produce polymers. Hydrocarbon sources from agriculture and other wastes may prove to be an important source of feedstock for polymer production. Pyrolysis extractions from organic waste and the fermentation from bacterial or small plants are other sources of producing simple organic chemicals.

Relative to Heat Reaction

All polymers are either thermoplastic or thermosetting. Probably one of the earliest distinctions between polymers was based on their reaction to heating and cooling.

Thermoplastics

Polymers that can be made to flow when heated and become solid when cooled are called **thermoplastics**. These materials may be softened repeatedly by heat and shaped into useful products. Most thermoplastic materials, including scrap or damaged pieces, may be recycled. These materials do not liquefy when heated but are in a highly viscous state. Continued heating above their *melting points* will cause them to degrade. Common analogies of ice, wax, or butter are generally given but do not truly represent the heated state of polymers.

The most useful members of the thermoplastic group are acrylics, cellulosics, polyamide, polystyrene, polyethylene, fluoroplastics, polyvinyls, polycarbonate, and polysulfone. Thermoplastic materials come in a variety of available forms and are generally fully polymerized. (See Forms and Polymerization.)

Thermosetting

Thermosetting materials cannot be reshaped or reformed once the material is set into a final structural framework. The heating and forming process causes them to undergo a *curing* reaction. Thermosets will char, burn, or degrade by continued heating, but they do not remelt. It is the resin monomer that determines whether the material will be thermoplastic or thermosetting.

Common analogies are baking a cake or boiling an egg. Elastomers may be thermoplastic or thermosetting materials. Many of the natural polymers were thermoplastic. This limited their usefulness; for example, natural rubber would become sticky in hot environments, until Goodyear discovered how to convert the polymer to a thermosetting material. Thermosetting materials are particularly useful in producing many composite materials because they are available in a variety of forms. Thermosetting resins in a partially polymerized liquid state may facilitate penetration and wetting of the other constituents.

Members of the thermosetting group include aminos, casein, epoxies, phenolics, polyester, silicones, and polyurethanes.

Relative to Polymerization Reaction

Molecules are formed or produced into polymers by two types of reactions: (1) addition and (2) condensation. You will recall that **poly** means many and **mer** means unit or part; thus, a polymer is made of many repeated units in a pattern. This repeating unit is the **mer** and should be considered to be bifunctional or difunctional (two reactive bonding sites). In the following example of polyvinyl chloride (PVC), the mer unit is shown:

```
      H  Cl  H  Cl  H  Cl  H  Cl  H  Cl
      |  |   |  |   |  |   |  |   |  |
· · ·—C—C —C—C —C—C —C—C —C—C —· · ·
      |  |   |  |   |  |   |  |   |  |
      H  H   H  H   H  H   H  H   H  H
                    →|mer|←
```

The number of times this mer unit is repeated in the polymer is called the *degree of polymerization* (DP). A subscript n is used to denote the number of like units:

$$
\left[
\begin{array}{cc}
H & Cl \\
| & | \\
C & -C \\
| & | \\
H & H
\end{array}
\right]_n
$$

where

$\quad$ H $=$ hydrogen atom

$\quad$ Cl $=$ chloride atom

$\quad$ C $=$ carbon atom

$\quad$ n $=$ degree of polymerization or DP

$\quad$ [ ] $=$ mer

$\quad$ __ $=$ bond

$\quad\quad$ The carbon atom backbone may have a number of different atoms or functional molecular groups attached to it. In Figure 2-1 the number of growing chain lengths is shown from a molecule of methane gas to a solid polyethylene plastics.

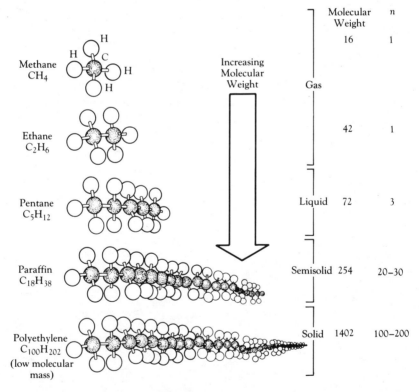

Figure 2-1. Growing chain length (molecular weight) from a single molecule of gas to a solid polymer.

Mer units cannot stand alone because of the requirement of the carbon's four bonding sites. Monomers (meaning a single unit or mer) are molecules: They can exist alone.

In the case of PVC, the monomer vinyl chloride (C_2H_3Cl) is shown as

$$
\begin{array}{ccc}
H & & Cl \\
| & & | \\
C & = & C \\
| & & | \\
H & & H
\end{array}
$$

The process of joining many monomers to each other in a repeating chain is called polymerization (meaning many mers joined). If the polymer consists of similar repeating units, it is known as a homopolymer (from **hom** or **homo**, meaning same). It is the repeating unit that distinguishes one polymer from another. When two or three different monomers are polymerized, the polymer is known as a copolymer or a terpolymer. Thus, if a number of vinyl chloride monomers are joined in a chain, the reaction is called **addition polymerization**. All the unsaturated (meaning capable of joining more) monomers are used to form the polymer:

$$
\left(
\begin{array}{ccc}
H & & Cl \\
| & & | \\
C & = & C \\
| & & | \\
H & & H
\end{array}
\right)_x
\longrightarrow
\left[
\begin{array}{ccc}
H & & Cl \\
| & & | \\
C & - & C \\
| & & | \\
H & & H
\end{array}
\right]_n
$$

The length of the chain has a profound effect on properties and molecular weight (mass). Molecular weight is a major factor in how easily the polymer is molded or shaped. Molecular weight refers to the average weight (M_w) of all the molecules in the mixture of different sized molecules in the polymer. As a rule, the higher the M_w, the tougher and more chemical resistant is the polymer. Of course, there is a practical limit to increasing M_w. An increased M_w means increased melt temperatures and viscosity. This slows processing, thus increasing costs and possible polymer degradation.

High molecular weight, high-density polyethylenes (HDPE) are linear polymers with molecular weights ranging from 200,000 to 500,000. This material has properties that differ significantly from low molecular weight polyethylene. For the materials ranging from

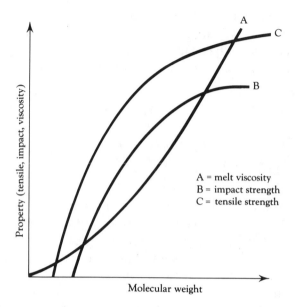

Figure 2-2. Property changes with increasing molecular weight (mass).

methane gas to polyethylene plastics, there is initially a great change in properties with increasing molecular weight. This is shown in Figure 2-2. Most polymers do not have extremely high average molecular weights. Molecular weight averages for selected polymers are shown in Table 2-1. The molecular weights of most polymers range from 5000 to 100,000.

Vinyl chloride monomers consist of two carbon atoms, three hydrogen atoms, and one chlorine atom. If there are 500 mers per

Table 2-1. Average
Molecular Weights of
Selected Polymers

Polymer	Molecular Weight
Paraffin	800
LDPE	15,000
HDPE	20,000
UHDPE	40,000
PVC	40,000
PP	40,000
PETP	25,000

polymer, the molecular weight is therefore

$$\left(\begin{array}{cc} H & Cl \\ | & | \\ C & = C \\ | & | \\ H & H \end{array} \right)_x$$

$$\frac{\text{weight}}{\text{mer}} = \frac{\text{g/mer weight}}{\text{mers/mer weight}}$$

$$= \frac{(2)\,(12) + (3)\,(1) = 35.5}{6.02 \times 10^{23}}$$

$$\text{molecular weight} = (\text{g/mer weight})\,(\text{mers/mole})$$

$$= (62.5)\,(500)$$

$$= 3.125 \times 10^4 \text{ g/molecular weight}$$

This means that 6.02×10^{23} (Avogadro's number) of these molecules weigh 31,250 g or roughly 48 lb.

The number of repeating molecule units, that is, degree of polymerization, is determined by dividing the molecular weight by the mer weight:

$$DP = \frac{\text{molecular weight}}{\text{mer weight}} \quad \text{or} \quad \frac{\text{mers}}{\text{mole}}$$

$$= \frac{31\ 252}{62.5}$$

$$= 500$$

or

$$\text{molecular weight of polymer} = (DP) \times (\text{molecular weight of monomer})$$

Not all the resulting polyvinyl molecules are identical. Since the polymer contains a mixture or range of molecular sizes, the average degree of polymerization is used to help determine the true properties for the polymer.

The water does not act as a solvent but suspends the monomer during the reaction. Relatively high molecular weight polymers are produced as the reaction is agitated and the water is removed. Table 2-2 shows polymerization methods for selected polymers. Many polymers may be polymerized by several polymerization methods. Poly-

Table 2-2. Polymerization Methods for Selected Polymers

Polymer	Additional Methods
Chloroprene (CR)	Emulsion
Low-density polyethylene (LDPE)	Bulk
High-density polyethylene (HDPE)	Solution
Polyacetals	Solution
Polyamide (PA)	Bulk, suspension
Polycarbonate (PC)	Bulk
Polyethylene terephthalate (PETP)	Bulk
Polyisoprene (IR)	Solution
Polymethyl methacrylate (PMMA)	Bulk, suspension, solution
Polypropylene (PP)	Solution
Polystyrene (PS)	Emulsion, suspension, bulk, solution
Polysulfides	Suspension
Polytetrafluoroethylene (PTFE)	Suspension
Polyvinyl chloride (PVC)	Emulsion, suspension
Polyvinyl acetate	Emulsion
Styrene–butadiene rubber (SBR)	Emulsion

styrene, for example, can be made by any of the four addition polymerization methods, while polyamide may be produced using condensation polymerization.

There are several addition polymerization practices beyond the scope of this book; consequently, they are only briefly summarized. Bulk or mass polymerization takes place when one or more monomers are joined by the addition of catalysts and initiators. The high-yield reaction produces the highest-purity polymer. Excellent optical properties of clarity and color are further advantages. Cast epoxy and methyl methacrylate are familiar examples of this process.

In solution polymerization, the monomer, solvents, initiators, and catalyst are heated together; therefore, recovery and handling of solvents add to the expense. Polymerization continues as the solvent is removed. Because the resulting polymer contains trace quantities of each solvent, optical properties do not equal those of bulk polymerization. These polymers are used in coating and impregnating applications, where the solvent is allowed to evaporate.

Emulsion polymerization uses water rather than an organic solvent to distribute energy, the emulsifier, and initiators. Water-based polymer adhesives and paints are examples of this practice. This process produces the highest molecular weight polymers. Suspen-

Table 2-3. Selected
Condensation Polymers

Phenol–formaldehyde
Polyacetal
Polyamide

Polyamide–imide
Polybutylene terephthalate
Polycarbonate

Polyester
Polyether
Polyimides

Polysiloxane (silicones)
Polysulfides
Polyurea

Polyurethane
Urea–formaldehyde

sion, bead, or pearl polymerization are names used to describe the technique of suspending monomer droplets in water.

Condensation polymerization causes monomers to join but a nonpolymerizable molecule by-product is formed. This residue or by-product is commonly water (H_2O). In condensation polymerization, the repeating molecules in the formed chain are different from the starting materials. Polyamides, polyesters, alkyds, and phenolics are formed by this method. It is important to note that condensation polymerization can lead to either thermoplastic or thermosetting polymers. Phenol–formaldehyde is produced when polymerization occurs between formaldehyde (CH_2O) and phenol (C_6H_5OH) and benzene. A water molecule is left as condensation. Selected condensation polymers are shown in Table 2-3.

Relative to Molecular Structure

All properties of the polymer are affected by the arrangement of the repeating units. Many polymers are composed of different repeating units with different structures. There are three major molecular structural categories: (1) linear, (2) branched, and (3) cross-linked or network.

If bifunctional monomers of PVC, PMMA, or PE are linked to form a long chain, the polymer structure is **linear**. These long molecules are twisted and tangled, something similar to a bowl of spaghetti.

The primary bonds (mostly intermolecular van der Waals) are weak. These weaker bonding forces allow the molecular chains to slide by each other and be deformed when heated. Linear, low-density polyethylene is much easier to form than high-density polyethylene. As a rule, most linear polymers are thermoplastic.

Not all linear molecules are homopolymers. When different monomer units are randomly placed in the linear chain, the polymer is called a **copolymer** or **random copolymer**. These random units have no definite order. The letters A and B may be used to help illustrate the formation of different monomers:

ABBABAAAABABBBAABBAAB

Alternating copolymers have monomers with definite ordered alteration:

ABABABABABABABABABABABABABABABABABABABAB

When long continuous blocks of different monomers are combined in the chain, the result is a **block copolymer**:

AAAAAAAAAAAAAAAAAAAABBBBBBBBBBBBBBBBBBBB

Many homopolymers and older polymers have been improved by copolymerization. The idea that we may want to attach other monomers or molecules in a random fashion along the linear chains is called **branched** or **graft** copolymerization. These side chains permit greater interlocking of the structure: As a result, deformation is more difficult. Some branching occurs naturally during polymerization and processing. The term graft copolymer is used to denote a controlled or planned action:

AAAAAAAAAAAAAAAAAAAAAA

| | |
B   B
B   B
B   B
B   B

Many polymers may be produced by causing chains to bond to each other during the polymerization process. A **cross-linked** or **network** polymer is one in which all the chains are connected together. These are primary, covalent bonds. It requires a great deal of energy to break these bonds. Cross-linked polymers are thermo-

complex. The ordered arrangement of the polymer chain into a crystalline structure is illustrated in Figure 2-3. The folded-chain or looped molecule alignment is called a **lamellar** structure. The fibrouslike aggregates of radiating crystals are called **spherulites**. Spherulites are present in most crystalline plastics and may range in diameter from a few tenths of a micrometer to several millimeters.

Crystallization raises the melting point, increases density, and generally improves mechanical properties. It lowers impact strength, solubility, and optical clarity. The crystalline content may exceed 95% with the remainder being an amorphous structure. Polymers possessing no crystallinity are termed amorphous (meaning without form). Because polymers are never 100% crystalline, they are sometimes called semicrystalline. For example, the crystallinity of LDPE is about 65%, while that of HDPE may exceed 95%.

Common crystalline polymers are PE, PTFE, PPA, and PA.

Molding processes greatly influence crystallinity. When PE is molten, the molecular chains are no longer aligned. Consequently, the melt is amorphous. Rapid quenching may prevent a large number of the chains from realigning to form crystals. This is sometimes done in the production of PET and PE containers to improve transparency. Processing may help align or orient the molecular chains. Polystyrene may form crystalline structures in narrow restrictions or molds. These areas will have a cloudy or milky appearance from crystal formation. The benzene side groups normally prevent chain alignment or significant crystal formation. Processing may cause molecular alignment during the polymer fiber production, and although this will affect optical properties, oriented alignment of molecules on one axis may be desirable.

If a polymer has crystalline alignment, light can be refracted and scattered. Therefore, LDPE should have greater clarity than HDPE because it has less crystallinity. On the other hand, polystyrene is amorphous and while it is clear it is also more easily attacked by solvents.

The hydrogen atom is rather small when compared with chlorine, fluorine, methyl, and benzene. Relative sizes of these side groups are shown in Figure 2-4. The physical size and bonding angle results in stereoisomerism of polypropylene (meaning three arrangement of mers). **Atactic** pertains to an arrangement that is more or less random, while **isotactic** pertains to a structure containing a sequence of regularly spaced asymmetric atoms arranged in like configuration

setting. If all the chains are cross-linked, the resulting "giant
ecule cannot be reshaped or formed without thermal decompos
or physical destruction of the molecule.

There are numerous additional structural arrangements that
ter the behavior of the polymer.

The concept of alloying compatible families of polymers togeth
is relatively new. There have been commercial alloys of ABS/PVC
PVC/PMMA, PMMA/PS, ABS/PC, and others for more than a decade.

A **plastics alloy** is formed when two or more different polymers
are physically mixed together during the melt. Although there may
be some bond attachment between chains, these should not be considered copolymers. Some authors refer to this combination as a
composite polymer (alloy).

The term **polyblend** is sometimes used to refer to plastics that
are modified by the addition of elastomers. Elastomer-modified PS,
PVC, PMMA, and ABS are commercially available. A plastics alloy or
polyblend may appear as

BBBBBBBBBBBAAAAAAA BBBBBBBBBBBBBBBBB
AAAAAAAAAAAAAAAAAAAAAAAAA BBBBB
BBBBBBBBBBBBBBBBBBBBBBBBBBBBB BBBBBBBBB

Another method used to strengthen the polymer is to add other
molecules periodically along the chain. This produces stronger attractive forces between chains and restricts chain movement. If we
take the uniform structure of polyethylene and replace some of the
hydrogen bonds with chlorine (Cl), methyl group (CH_3), benzene
(C_6H_6), or fluorine (F), the polymers PVC, PP, PS, and PTTE will be
formed. Each of these polymers has uniquely different properties. A
new family of polymers does not have to be developed. Even the
substitution of a limited number of these side groups will effect a
change in the polymer.

Relative to Crystal Structure

The polyethylene chain is regular with no side groups. This allows
the molecular chains to pack into a crystal lattice. This implies that
a large number of molecules are easily positioned because of the
weak van der Waals bonds between the linear chains.

Crystallinity is the three-dimensional arrangement of molecules
in a regular pattern. The crystalline morphologies of polymers are

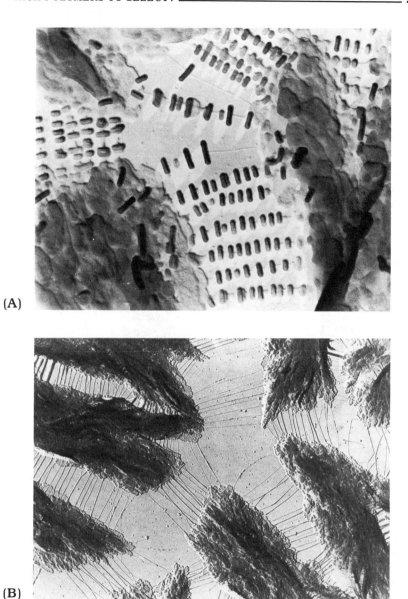

(A)

(B)

Figure 2-3. (A) Lamellar crystals of polyethylene were formed by depositing polymer from solution on links. (B) The first direct observation of links between the crystals in a polymer was made by scientists at Bell Laboratories. Fine fibrous links bridge the gaps between the radial arms of a single spherilite of polyethylene. (Magnification 10,000 diameters.) (AT&T Bell Laboratories)

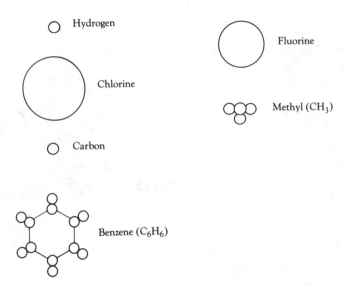

Figure 2-4. Relative sizes of selected atoms and side groups.

in a polymer chain. **Syndotactic** pertains to a polymer molecule in which groups of atoms that are not part of the primary backbone structure alternate regularly on opposite sides of the chain. In the atactic structure of PP, the methyl groups make it difficult for chains to align or form crystals. This polymer is soft, weak, and less cloudy than the more crystalline isotactic and syndotactic forms.

Thermoplastic Polymers

This brief discussion of thermoplastic polymers is intended to acquaint the reader with available materials. The use of the data provided is not to be taken as a guide to the selection of any material. New, more current data are available from a number of sources. The selection of the matrix in composite design must include careful consideration of the material selection, design configuration, processing, and materials manufacturer. (See Design Practice.)

As a rule, most homopolymer thermoplastics are furnished chemically complete or polymerized. They are available in a number of available forms. **Pellet preforms** are widely used because they may be compounded to processor specifications. **Adhesive forms** are available in a number of physical forms: powders, films, dispersions, hot melts, pastes, liquids, and two-part components. Thermoplastic polymers also come in a number of **dimensional** or **profile forms,** with sheets, films, rods, and tubes being typical examples. Synthetic **fibers** and **filaments** continue to grow in importance. **Casting** and **coating** formulations are other well-known forms. (See Casting and Coating Processes.) We have all used **cellular** or **foamed** plastics and elastomers as containers, insulations, and shoe soles.

Most thermoplastics may be compounded to improve properties. Homopolymers generally have low service temperatures and suffer from creep.

Distinguishing characteristics of individual groups of thermo-plastic materials are arranged in alphabetical order. Only important alloys/blends are included. It should be apparent that there may be no end to the combination of alloying/blending of polymeric materials. As defined earlier, alloys are mixtures of two or more polymers. Production methods vary from simple mechanical melt blending to complex interaction of either miscible (meaning capable of mixing without separation) or immiscible (incapable of mixing) mixtures. Some use the terms **alloy** and **blend** interchangeably. In this discussion the term **blend** is used only as a verb. To be classed as an alloy, at least two of the polymer components must be in concentrations greater than 5%.

Acetal Copolymer

Acetal copolymer materials are similar in performance and applications to many grades of acetal homopolymers. The high thermal stability and chemical resistance are attributed to the carbon-to-carbon links in the polymer chain and the addition of hydroxyethyl terminal units. Copolymerization of trioxane with small amounts of selected comonomer is randomly distributed on the polymer chain. This structure provides good creep resistance. Improvements in dimensional stability, high flexural modulus, and low warpage are provided by filled grades (see Table 2-4*). Silicone oil, glass, and fluoroplastic fibers are added to the matrix for selected bearing applications.

Processing parameters are similar to some homopolymer grades. They are injection-blow, extrusion-blow, and injection molded.

*In Tables 2-4–2-28, the following letters are used for the respective grades: E = extrusion; I = injection; C = compression; T = transfer. Also in these tables, the following units are used: molding pressure, psi; tensile strength, psi; compressive strength, psi; flexural strength, psi.

Table 2-4. Selected Properties of Acetal

Property	20% Glass	25% Glass Coupled Copolymer
Relative density	1.56	1.59–1.61
Melting temperature (°C) T_m	181	175
T_g	—	—
Processing range (°F)	I:350–480	I:380–480
Molding pressure	10–20	10–20
Shrinkage	0.009–0.012	0.004 (flow)
		0.018 (trans)
Tensile strength	8,500	16,000–18,500
Compressive strength	18,000 at 10%	17,000 at 10%
Flexural strength	15,000	18,000–28,000
Izod impact (ft-lb/in.)		
(1/8 in. specimen)	0.8	1.0–1.8
Linear expansion		
(10^{-6} in./in./°C)	36–81	44
Hardness, Rockwell	M90	M79, R110
Flammability	HB	HB
Water absorption, 24 h (%)	0.25	0.29

Acetal copolymers find applications in plumbing valves, seat belt components, aerosol bottles, appliance parts, and chain links.

Acetal Homopolymer (POM)

This crystalline polymer is known as polyacetal, polyformaldehyde, and polyoxymethylene (POM). This latter term is best because of the carbon-to-oxygen bond rather than the more typical carbon-to-carbon bond. Homopolymers have high tensile, compressive, and shear strengths. In addition, the material has a high flexural fatigue limit and excellent impact resistance. POM is competitive in cost and performance with many nonferrous metals. It compares and competes with polyamides in many applications. Acetal homopolymers are resistant to a wide variety of solvents, acids, and bases. However, exposure to ultraviolet light may cause surface chalking, reduced molecular mass, and slow degradation. Homopolymers have a maximum continuous use temperature of 185°F (85°C). The matrix has low coefficients of friction but may be filled with fluoropolymer fibers for improved abrasion resistance and even lower coefficients of friction. Stiffer, more warp- and creep-resistant grades are glass. Alloying butadiene or other elastomer monomers results in tough polymers.

Acetal homopolymers are injection, blow, and extrusion molded. Adequate ventilation during molding is recommended because acetals release a potentially toxic gas during degradation.

Typical applications include carburetor floats, latches, lock parts, gears, springs, shower heads, sink sprayers, appliance handles, fan blades, key buttons, disposable lighter bodies, zippers, and bushings.

Acrylic

Acrylics are a family of polymeric materials. The term **acrylic** includes acrylic and methacrylic esters, acids, and other derivatives. The basic acrylic monomer is shown in Figure 2-5 with two possible radical replacement sites. (See also Polyacrylates, Polyacrylonitrile, Polyethyacrylonitrile, and ABS.)

Acrylic monomers are polymerized by bulk, solution, emulsion, and suspension. In all cases, an organic peroxide catalyst is used to begin polymerization methods.

Polymethyl methacrylate (PMMA) is an atactic, amorphous, thermoplastic material. It comes in a number of available forms, including powders, monomer resins, and dimensional shapes. Resins are used to produce cast sheets or other shapes. (See Casting Process.) Sheets may also be produced by continuous casting or by extrusion. There is considerable shrinkage in polymerization; consequently, mold designs must accommodate these tolerance differences. Pellet formulations are molded by all thermoplastic molding techniques.

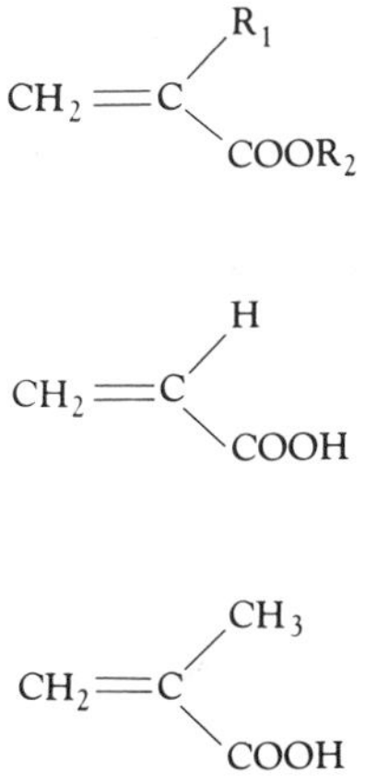

$$CH_2 = C \bigg\langle {}^{R_1}_{COOR_2}$$

$$CH_2 = C \bigg\langle {}^{H}_{COOH}$$

$$CH_2 = C \bigg\langle {}^{CH_3}_{COOH}$$

Figure 2-5. Acrylic formula, with two possible radical replacements.

It is the wide variety of properties, available forms, processing techniques, and copolymerization options that make this polymer popular. Well-known trade names for PMMA are Plexiglas, Lucite, and Acrylite.

PMMA polymers have excellent optical, thermal, electrical, and chemical properties (see Table 2-5). They have poor resistance to ketones and chlorinated solvents. Most grades may be exposed to service temperatures of 200°F (93°C). Outstanding weatherability makes acrylics an excellent choice for exterior applications. Artists and homeowners know the merits of acrylic polymers.

Table 2-5. Selected Properties of Acrylic

Property	Acrylic–PVC Alloy	PMMA	Heat-Resistant
Relative density	1.11	1.17–1.20	1.16–1.19
Melting temperature (°C) T_m	—	—	—
T_g	1.5	90–105	100–125
Processing range (°F)	—	C:300–425 I:325–500 E:360–507	C:350–450 I:400–500 E:360–507
Molding pressure	—	5–20	6–30
Shrinkage	—	0.001–0.004 (flow) 0.002–0.008 (trans)	0.003–0.008
Tensile strength	6,500	7,000–11,000	9,300–10,000
Compressive strength	8,400	10,500–18,000	15,000–17,000
Flexural strength	10,700	10,500–19,000	12,000–18,000
Izod impact (ft-lb/in.) (1/8 in. specimen)	15	0.3–0.6	0.3–0.4
Linear expansion (10^{-6} in./in./°C)	—	50–90	50–71
Hardness, Rockwell	R99–105	M68–105	M94–105
Flammability	HB	HB	HB
Water absorption, 24 h (%)	0.06	0.1–0.4	0.2–0.3

Automobile taillight lenses, signs, glazing, marine windshields, aircraft canopies, and architectural panel fascia are common uses.

Composite structures have been produced for space applications using a graphite and acrylic matrix. Loading of woven graphite cloth to 65% has been used to produce a laminate with impressive properties, including excellent wetting, low forming temperature, and low cost, which make it a possible candidate for beam construction in space.

Glass-reinforced sheets are used to produce sanitary ware, vanities, tubs, and counters. Gel coats, protective liquid coatings, or clear films may be used with reinforcements as cover stock to produce a laminate or clad over metal or other substances. Heavily filled and reinforced resins formulated to cross-link into a thermosetting matrix are also used to produce marblelike bathroom fixtures and furniture.

An alloy of acrylic and polyvinyl chloride (PMMA/PVC) is produced into tough, durable sheet. The smooth, glossy finish may be attributed to the acrylic and the durability to PVC. The material may be formed mechanically at ambient temperature but has a useful service temperature of less than 150°F (66°C).

Acrylic–Styrene–Acrylonitrile (ASA)

The outstanding characteristic of this terpolymer is UV resistance. The acrylic content makes it an ideal choice for outdoor applications. Like many terpolymers, varying the percentages of each component may enhance a specific property; for example, the addition of PVC to the compound increases strength and makes it more resistant to selected chemicals and provides a high flammability rating.

The polymer matrix is hygroscopic but may be processed by all thermoplastic equipment. The excellent surface gloss makes it attractive as a cover layer for coextrusion over ABS, PC, or PVC.

Exterior applications include signs, downspouts, gutters, ATV bodies, camper tops, and house siding.

Acrylonitrile–Butadiene–Styrene (ABS)

This is a graft terpolymer of three different monomers: acrylonitrile, butadiene, and styrene. It is produced by emulsion, bulk, and suspension polymerization. The amorphous material may be compounded to meet specific properties by changing the component ratios. Increasing the rubber content increases impact strength but lowers tensile strength and surface gloss. It is available in pellet, powder, sheet, and foam forms. Because it is slightly hygroscopic, all grades must be dried before processing.

These materials may be fabricated by all thermoplastic processing techniques. Reinforced and filled grades have high melt viscosities. General fillets, streamlining, and full round runners may be needed. Processing temperatures vary by grade. (See Table 2-6.)

ABS materials are characterized as tough, low cost, easily fabricated and processed materials. They have been used for pipes, power tool housings, and appliance components. Refrigerator tanks, door liners, and business equipment housings have been the fastest growth areas. Reinforced and filled grades are used for radiation shielding (electromagnetic interference). Glass fiber and aluminum particles to more than 40% loading have been used.

Table 2-6. Selected Properties of ABS

Properties	ABS/PVC	20% Glass Reinforced	20% PAN Carbon Fiber	40% Aluminum Flake
Relative density	1.20–1.21	1.18–1.22	1.13–1.14	1.61
Melting temperature ($^\circ$C) T_m	—	—	—	—
T_g	—	100–110	100–110	—
Processing range ($^\circ$F)	—	C:350–500 I:350–500	I:430–500	I:450–550
Molding pressure	—	15–30	15–30	—
Shrinkage	0.003–0.005	0.001–0.002	0.0005–0.003	0.001
Tensile strength	5,800	11,000–13,000	15,000–16,000	4,200
Compressive strength	—	13,000–14,000	17,000	6,500
Flexural strength	7,900–9,600	14,000–17,500	23,000–25,000	7,800
Izod impact (ft-lb/ in.) (1/8 in. specimen)	6.5–10.5	1.2–1.4	1.0	2.0
Linear expansion (10^{-6} in./in./$^\circ$C)	46084	20–21	18	40
Hardness, Rockwell	R100–106	M85–R107	R108	R107
Flammability	V-0	HB	—	V-1
Water absorption, 24 h (%)	—	0.18	0.17	0.23

Acrylonitrile–Chlorinated Polyethylene–Styrene (ACS)

This terpolymer is similar in properties to ABS. Because of the chlorine content, it surpasses ABS in flame-retardant properties, weatherability, and service temperatures. Processing temperatures must not exceed 428°F (220°C) to prevent degassing of the chlorine. This polymer may be reinforced or filled to improve selected properties. ACS may be processed using all thermoplastic processing equipment. The polymer is slightly hygroscopic.

Applications are similar to those of ABS. Most are used for office machine housings, appliance cases, and electrical connectors.

Cellulosics

The term **cellulosics** refers to a family of plastics derived from wood pulp, cotton linters, or other cellulose sources. The most commonly used cellulosic polymers are ethyl cellulose, cellulose acetate, cellulose acetate butyrate, and cellulose acetate propionate. Cellulose nitrate and cellulose triacetate are also produced and used. These materials are not thermoplastic and must be processed using solvent dispersion and recovery systems. Although these plastics are not cross-linked, they will degrade rather than soften and flow (see Table 2-7). Solvent cast films of cellulose triacetate are used for photographic film base, dielectric separators, and tape.

Ethyl cellulose (EC) is produced from alkali cellulose and ethyl chloride to form ethyl cellulose. Nearly half of the polymer may be composed of ethyl groups. The polymer is light amber and may come in a number of formulations. This tough, flexible, and moisture-resistant plastics may be filled and reinforced.

Typical applications may include football helmets, flashlight cases, furniture trim, and blister packages.

Cellulose acetate (CA) is produced by the removal of some of the acetyl groups from cellulose triacetate. Cellulose acetates are made into fibers, coatings, molded brush handles, combs, eyeglass frames, and transparent packages.

Cellulose acetate butyrate (CAB) is produced from the reaction of butyric acid and acetic anhydride. Familiar applications are screwdriver handles, pen and pencil barrels, blister packages, and some display signs. The material is used as film layers with metal, foils, papers, and other plastics in special packaging applications.

Cellulose acetate propionate (CAP) is formed like other acetates

Table 2-7. Selected Properties of Cellulosics

Properties	Ethyl Cellulose	Cellulose Acetate	Cellulose Acetate Butyrate
Relative density	1.09–1.17	1.22–1.34	1.15–1.22
Melting temperature			
($°C$) T_m	—	—	140
T_g	135	230	—
Processing range (°F)	C:250–390	C:260–420	C:265–390
	I:350–500	I:335–490	I:335–480
Molding pressure	8–32	8–32	8–32
Shrinkage	0.005–0.009	0.003–0.010	0.003–0.009
Tensile strength	2,000–8,000	1,900–9,000	2,600–6,900
Compessive strength	—	3,000–8,000	2,100–7,500
Flexural strength	4,000–	2,000–	
	12,000	16,000	1,800–9,300
Izod impact (ft-lb/in.)			
(1/8 in. specimen)	0.4	1.0–7.8	1.0–10.9
Linear expansion			
(10^{-6} in./in./°C)	100–200	80–180	110–170
Hardness, Rockwell	R50–115	R34–125	R31–116
Flammability	HB	HB	HB
Water absorption, 24 h			
(%)	0.8–1.8	1.7–6.5	0.9–2.2

with the addition of propionic acid. These propionyl groups may exceed 45% of the polymer. This plastic is commonly thermoformed and molded. It must compete with CAB which has similar product applications.

Ethylene Acid

By varying the pendent carboxyl groups on the polyethylene chain, a branched, random copolymer is produced. A wide variety of properties may be produced by changing the proportion of acid groups. They have properties similar to LDPE. As a rule, they have outstanding toughness, adhesive qualities, and stress-crack resistance.

They may be processed like LDPE but corrosion-resistant processing components should be used. Molds must be designed to accommodate the adhesive qualities of the copolymer.

Coextruded films and laminates on a number of substrates are important applications. Composite aluminum foil pouches are used for towelettes and dispensing a number of products. Composite structures of polymer, foil, wood, glass reinforcements, and other constituents are used in selected automotive, aircraft, and building components.

Ethylene–Ethyl Acrylate (EEA)

A variety of properties are produced in this copolymer by varying the ethyl acrylate pendent groups in the ethylene chain. Properties vary from rubbery to tough polyethylenelike polymers, depending on the amount of increase in the molecular weight. Processing and applications are similar to EMA.

Ethylene–Methyl Acrylate (EMA)

This olefin copolymer may be composed of 40% by weight of methyl acrylate. The elastomeric characteristics and compatibility with all polyolefins allow for a number of alloying techniques and a diversity of properties. The polymer is easily processed and may be filled to more than 60% without loss of all rubberlike properties. There is good adhesion to most substrates in extrusion coatings or laminates.

EMA copolymers are used as disposable medical gloves and tough, heat-sealable layers for numerous composite packaging applications.

Ethylene–Vinyl Acetate (EVA)

When 5–50% by weight of vinyl acetate monomers are added to the polymer chain, a copolymer similar to EEA is produced. If the vinyl acetate side groups exceed 50%, they are considered vinyl acetate–ethylene (VAE) copolymers. Solvent cast films of cellulose triacetate are used for photographic film base, dielectric separators, and tape.

Fluoroplastics

Fluoroplastics are a family of plastics with an alkenelike structure with some or all of the hydrogen atoms replaced by fluorine. The term **fluorocarbon** is sometimes incorrectly used to describe this family of materials. Fluorocarbon should be used to refer to compounds containing fluorine and carbon.

Important members of the fluoroplastic family include ethylene–chlorotrifluoroethylene (ECTFE), ethylene–tetrafluoroethylene

(ETFE), fluorinated ethylene propylene (FEP), perfluoroalkoxy (PFA), polychlorotrifluoroethylene (PCTFE), polytetrafluoroethylene (PTFE), polyvinylfluoride (PVF), and polyvinylidene fluoride (PVDF).

As a family of materials, they are known for their wide temperature range, inertness to acids, alkalis, and solvents, and excellent electrical properties (see Table 2-8).

Polytetrafluoroethylene accounts for nearly 90% of the fluorinated plastics used.

Carbon-to-fluorine bonds are stronger than carbon-to-hydrogen or carbon-to-chloride bonds. Densities range from 1.78 to 2.2. It is the fluorine bonds that provide this family with excellent thermal stability, resistance to solvents, and outstanding electrical proper-

Table 2-8. Selected Properties of Fluoroplastics

Properties	Polytetrafluoroethylene 25% Glass Fiber Reinforced	Fluorinated Ethylene Propylene 20% Milled Glass Fiber	Polyvinylidene Fluoride 30% PAN Carbon Fiber
Relative density	2.2–2.3	—	1.74
Melting temperature (°C) T_m	327	—	—
T_g	—	262	—
Processing range (°F)	—	I:600–700	I:430–500
Molding pressure	3–8	10–20	—
Shrinkage	0.018–0.020	0.006–0.010	0.001
Tensile strength	2,000–2,700	2,400	14,000
Flexural strength	2,000	4,000	19,800
Izod impact (ft-lb/in.) (1/8 in. specimen)	2.7	3.2	1.5
Linear expansion (10^{-6} in./in./°C)	77–100	22	—
Flammability	V-0	V-0	V-0
Water absorption, 24 h (%)	—	0.01	0.12

ties. Fluoroplastics have low coefficients of friction and may be used at temperatures from 392 to 500°F (200 to 260°C). Processing all fluoroplastics requires special care to prevent thermal decomposition. These fumes are noxious and toxic. Only corrosion-resistant processing surfaces should come in contact with the polymer.

Ethylene–Chlorotrifluoroethylene (ECTFE)

This linear chained copolymer consists of alternating ethylene and chlorotrifluoroethylene repeating units. It has high performance properties common to other fluoroplastics. It surpasses them in impermeability, tensile strength, wear resistance, and creep. ECTFE may be extrusion processed into foam-coated wires and cables. Films find special use as release agents, tank linings, and dielectrics.

Ethylene–Tetrafluoroethylene (ETFE)

The backbone of this copolymer is composed of alternating units of ethylene and tetrafluoroethylene. Processing techniques, properties, and applications are similar to those of ECTFE.

Fluorinated Ethylene Propylene (FEP)

Polyfluoroethylenepropylene is a copolymer with CF_3 groups. This allows a crystalline melt at about 554°F (290°C), thus allowing the polymer to be processed with conventional processing equipment. The polymer is relatively soft but has low coefficients of friction and is used for lining chutes, pipes, and coatings and is molded into gears, impellers, printed circuit boards, bearings, and numerous wire coatings. Fillers are used as mold-releasing agents in composite laminating and bag molding operations.

Perfluoroalkoxy (PFA)

This plastics has properties and applications similar to PTFE and FEP. It can be used in temperatures up to 500°F (260°C). PFA polymers may be molded by conventional processing techniques.

Polychlorotrifluoroethylene (PCTFE)

This crystalline polymer with a melting point of 424°F (218°C) is more difficult to process because of its high melt viscosity and potential to degrade. The presence of the chlorine atoms in the molecular chain allow this polymer to be attacked and dissolved in selected chemicals. It possesses outstanding barrier properties against water vapor and gases. Transparent films are used in pharmaceutical packaging, while films and sheets are fabricated into linings for pipe and chemical processing equipment. Polychlorotrifluoroethylene is harder, more flexible, and more expensive than PTFE.

Polytetrafluoroethylene (PTFE)

Polytetrafluoroethylene is a highly crystalline, waxy thermoplastic. Strong fluorine bonds and interlocking atoms prevent conventional processing. Most products are formed from powders or water-based dispersions. Compressed or extruded shapes are sintered at high temperatures, similar to those used in the metals or ceramics industries. Dispersions are used to coat home cookware, tools, and other nonstick product applications. Glass, metals, carbon, and graphite are sometimes added. PTFE compounds are used as bushings, seals, bearing pads, and particle-sized lubricants. Films and tapes are cut or sliced from sheet stock and then used for release layers and dielectric insulation. Dry PTFE powders make excellent lubrication and releasing agents for lamination of composite structures. Residual particles are easily removed from the cured surfaces.

Polyvinyl Fluoride (PVF)

Polyvinyl fluoride is not as chemical or heat resistant as PTFE, PCTFE, FEP, or PFA, but it does have better melt processability. It is used for numerous film and coating applications. For example, coatings and films are applied to automotive parts, lawn mower housings, house shutters, gutters, metal siding, plywood, hard board, reinforced concrete fascia, wall coverings, glazing panels, metal foils, and other exterior coating applications.

Polyvinylidene Fluoride (PVDF)

This high molecular weight, crystalline polymer closely resembles polyvinyl fluoride. It is superior in strength, wear resistance, and creep resistance to PTFE, FEP, and PFA. The alternating CH_2 and CF_2 groups in the chain contribute to its tough, flexible characteristics. It has good weatherability and chemical resistance but may be solvent cemented. Applications are molded valves, tubing, ducting, and electronic insulation. A familiar use is the tough coating seen on aluminum and steel siding for homes.

Ionomer

This semicrystalline polymer contains random ionic cross-links along the polymer chain, and these weaker bonds are easily broken when heated. This high molecular weight material is made by polymerization of ethylene and methacrylic acid and may be processed using conventional thermoplastic techniques. Ionomers must compete with polyethylene since many of the properties are similar. Although they are more expensive than PE, ionomers possess a high molecular vapor permeability. Being less crystalline, ionomers can be molded and formed into transparent forms. They are tough, flexible, abrasion-resistant materials but have an upper service use of about 176°F (80°C). Creep and heat resistances are significantly increased when reinforced or filled.

Composite structures are a major market application for ionomers. Laminated or coextruded films are used as tear-open pouches for pharmaceutical and food packaging. Ionomer and metal foils are finding increasing applications in combination with other substrates as coatings and food pouches or containers. Heat-sealable skin and blister packaging of products continues to grow. Foamed applications have been used as bumper guards, resilient shoe components, wrestling mats, and ski lift seat pads. Ionomer coatings on golf balls and bowling pins extend the service life of these products.

Nitriles

Numerous formulations of copolymers with a nitrile functionality of over 50% are called nitrile polymers. Nitriles are compounds that contain the C≡N group. Because high nitrile-containing polymers possess a very low gas permeability, they are commonly called barrier polymers. Most combinations are based on acrylonitrile (AN) or methacrylonitrile. Some formulations may reach 75% AN. Acrylonitrile may be modified with monomers of styrene and/or butadiene. One formulation, for example, contains 65% AN, 6% butadiene, and 20% styrene. All formulations are amorphous, with a slight yellow tint, and are characterized by good chemical resistance and low gas and odor permeability.

Although heat sensitive, nitrile barrier polymers have been used with all thermoplastic processing techniques.

Major applications are in the food packaging industries. Containers for soft drinks, juices, dry soups, jellies, and other products continue to grow rapidly. The AN residual monomers of these food containers cannot exceed 0.10 ppm. (See Polyacrylonitrile–Styrene–Acrylonitrile, Acrylonitrile–Butadiene–Styrene.)

Polyallomers

Polyallomers are produced by a process called allomerism (meaning variability in chemical constitution without a change in crystalline form). It is conducted by alternately polymerizing ethylene and propylene monomers. The term **polyallomer** is used to distinguish this alternately segmented polymer from homopolymers and copolymers of ethylene and propylene. This highly crystalline block polymer has some of the properties of both polyethylene and polypropylene. Properties include high impact strength, flexibility, and low density (0.90).

Processing and applications are similar to PE and PP. Their flexibility has been used in integrally hinged boxes, looseleaf binders, and folders. They must compete with PE for use in food containers and film production.

Polyamide (PA)

Polyamide is probably the best known of all synthetic materials. The trade name Nylon has been associated with variations of polyamide for so long that **nylon** is now an acceptable generic name. Polyamide is a family of polymers formed by the condensation of diamines with dibasic acids or those from the polymerization of lactams and amino acids. By varying the diamines, dibasic acids, or lactams or dispersing other polymer segments on the polymer chain, an infinite number of polyamide polymers may be produced. These may range from low molecular weight, low melt viscosity crystalline polymers to high molecular weight, high melt viscosity amorphous materials. The addition of olefinic materials to the PA matrix results in a tougher, impact-resistant plastics. Copolymerization with polyester or polyester segments produces heat-resistant elastomers. Amino-containing formulations may result in cross-linked thermosetting materials.

Nylons are designated by the number of carbon atoms attached to both the acid and amine groups. Nylon 6/6, 6, 6/9, 6/10, 6/11, 6/12 and other combinations are produced. This structural difference accounts for the variations in melting points, stiffness, and water absorption (see Table 2-9). As a family of polymers, they are strong, tough, chemically resistant, and have good abrasion resistance. All are hygroscopic. Polyamides may be filled and reinforced to improve selected properties. Carbon fiber reinforced grades may permit modulus of elasticity values over 5×10^6 psi. Glass- and mineral-filled combinations impart stiffness, reduce warpage, and increase strength.

All thermoplastic processing equipment may be used. Polyamides must compete with acetals and fluoroplastics for many applications. Low coefficients of friction make them popular as gears in windshield

Table 2-9. Selected Properties of Polyamide

Properties	6/6 40% Glass and Mineral Reinforced	6/6 30% Graphite Fiber	6/6 40% Aluminum Flake	6/12 30–35% Glass Fiber Reinforced
Relative density	1.42–1.49	1.28–1.43	1.48	1.30–1.38
Melting temperature ($^\circ$C) T_m	255–260	265	265	215–217
T_g	—	—	—	—
Processing range ($^\circ$F)	I:510–590	I:500–575	I:525–600	I:450–550
Molding pressure	9–20	10–20	10–20	4–20
Shrinkage	0.002–0.005	0.002–0.003	0.005	0.003–0.005
Tensile strength	15,500–20,000	32,000–35,000	6,000	20,000–22,000
Compressive strength	18,000–37,000	27,000	7,500	22,000
Flexural strength	24,000–28,900	45,000–51,000	11,700	32,000
Izod impact (ft-lb/in.) (1/8 in. specimen)	0.6–1.1	1.5	2.5	1.8
Linear expansion (10^{-6} in./in./$^\circ$C)	20–29	11	22	—
Hardness, Rockwell	M95–98	R120	R114	M93;E40–50
Flammability	HB	—	—	V-0
Water absorption, 24 h (%)	0.4–0.9	0.5	1.1	0.2

wipers and speedometers. Films are used for boil- and bake-in-the-bag food containers. Other applications include gun stocks, bicycle wheels, brush bristles, sutures, fishing line, power steering reservoirs, steering column housings, and cable ties. Polyamide may be used as a protective coating on metal substrates. Hot melt adhesives and two-part polyamide–epoxy combinations are also used. Tough, RIM molded parts are growing in demand.

The importance and impact of polyamide fibers cannot be overstated. They began as a substitute for natural fibers. Today, the fibers are manufactured in a variety of compositions and properties. Aromatic ring structured polyamide fibers are not melt processable, so these polymers must be made by dry-set wet spinning techniques. These are important reinforcing fibers in conveyor belts and tires and composite structures. (See Reinforcements, Fibers.)

Polyamide–Imide

Polyamide–imide is a combination of the nitrogen bond in polyamide and the ring structure of aromatic diamines of polyimide. This opaque, condensation polymer has exceptional creep and impact resistances. It is characterized by good dimensional stability, good chemical resistance, and a continuous use temperature of more than 500°F (260°C) (see Table 2-10). There are glass, graphite, and

Table 2-10. Selected Properties of Polyamide–Imide

Properties	30% Glass Fiber Reinforced	Graphite Fiber Reinforced
Relative density	1.57	1.41
Melting temperature (°C) T_m	—	—
T_g	275	275
Processing range (°F)	C:630–650	C:630–650
	I:600–675	I:600–675
Molding pressure	15–40	15–40
Shrinkage	0.003–0.005	0.000–0.002
Tensile strength	28,300	29,800
Flexural strength	46,000	45,900
Izod impact (ft-lb/in.) (1/8 in. specimen)	2.0	1.2
Linear expansion (10^{-6} in./in./°C)	18	11
Hardness, Rockwell	E94	E94
Flammability	V-0	V-0

other filled grades available. When compounded with graphite or fluoroplastic powders, polyamide-imide can be molded into bearing materials with low coefficients of friction and wear.

Polyamide—imide may be melt processed into jet engine components, generator parts, bushings, seals, and parts for business machines. Continued growth is expected for applications in the aerospace and electronics industries.

Polyarylate

Polyarylate is a light amber-colored polymer made from iso- and terephthalic acid and bisphenol A. It is an aromatic polyester thermoplastic material. Polyarylate is known for its combination of clarity, heat deflection resistance, high impact strength, and excellent weatherability. Glass-reinforced grades with more than 40% loadings improve stiffness, deflection temperature, and tensile strength.

A number of alloy and filled grades are available but all must be dried before being processed by conventional thermoplastic equipment.

Polyarylate exterior applications include glazing, halogen lamp lenses, traffic signal components, and aircraft. Some are used for selected microwave cookware pieces. (See Aromatic Polyester and Polyester.)

Polyarylsulfone

In polyarylsulfone, the bisphenol groups are linked by ether and sulfone groups. The term **aryl** refers to a phenyl group derived from an aromatic compound. It offers properties similar to other aromatic sulfones. They offer long-term thermal stability and continuous use temperatures of over 356°F (180°C). The material is stiff with a flexural modulus of 400,000 psi. Reinforced and filled grades are available.

Polyarylsulfone must be dried before processing on conventional thermoplastic equipment.

Uses for polyarylsulfone include circuit boards, high-temperature bobbins, sight glasses, lamp housings, electrical connectors, and housings and panels of composite materials for numerous transportation components.

Polybutylene (PB)

Polybutylene is a flexible, semicrystalline, linear polyolefin. Molecular weights vary from 230,000 to over 750,000. There are five crystalline modifications of the butene monomer, offering a variety of property ranges. These materials exhibit a broad range of flexibility and are generally resistant to creep, environmental stress cracking, chemicals, and abrasion.

They may be processed using all thermoplastic techniques. The largest single application is for hot and cold water piping and tank liners. Polybutylene is used as a hot melt adhesive. It is coextruded and laminated to other films and used as moisture barriers, heat-sealable packages, compression wraps, and hot fill containers. (See Polyolefin, Polybutylene Terephthalate.)

Polybutylene Terephthalate (PBT)

Polybutylene terephthalate is the condensation product of dimethyl terephthalate and 1,4-butanediol. It is also called a polytertetramethylene terephthalate.

This polymer is sometimes referred to as a thermoplastic polyester. Molecular weight averages may range from 36,000 to 100,000. Glass- and mineral-filled or reinforced grades have good heat resistance, dimensional stability, low moisture absorption, and reduced warpage (see Table 2-11). Most products have a glossy surface appearance.

A variety of conventional thermoplastic techniques are used to mold distributor caps, rotors, power tool housings, iron and toaster housings, and numerous electronic parts.

Polycarbonate (PC)

Polycarbonate is a linear polyester because it contains an ester of carbonic acid and an aromatic bisphenol. Polycarbonate may be blended with other polymers for a variety of properties. This amorphous polymer has excellent impact strength, creep resistance, transparency, and moldability (see Table 2-12). The unique properties are due to the carbonate and rings in the molecular chain.

All grades are easily processed but hygroscopic. Glass fiber reinforcement improves creep resistance and dimensional stability.

Typical applications include beverage pitchers, mugs, food processor bowls, trays, safety helmets, pump impellers, grill opening

Table 2-11. Selected Properties of Polyester

Properties	PBT 30% Glass Fiber Reinforced	PBT 40–45% Glass Fiber and Mineral Reinforced	PBT 35% Glass Fiber and Mica Reinforced	PET Unfilled
Relative density	1.48–1.52	1.65–1.74	1.59–1.73	1.34–1.39
Melting temperature (°C) T_m	232–267	220–228	220–224	254–259
T_g	—	—	—	73
Processing range (°F)	1:440–530	1:450–500	1:480–510	1:540–600
Molding pressure	5–15	10–15	9–15	2–7
Shrinkage	0.002–0.008	0.003–0.010	0.003–0.012	0.020–0.025
Tensile strength	14,000–19,000	12,000–14,600	11,400–13,800	7,000–10,500
Compressive strength	18,000–23,500	21,000	19,000	11,000–15,000
Flexural strength	22,000–29,000	18,500–23,000	18,000–22,000	14,000–18,000
Izod impact (ft-lb/in.) (1/8 in. specimen)	0.9–1.6	0.7–1.7	1.3–1.8	0.25–0.65
Linear expansion (10^{-6} in./in./°C)	25	15	20	65
Hardness, Rockwell	M90	M75–86	M50	M94–101
Flammability	HB	HB	HB	HB
Water absorption, 24 h (%)	0.06–0.08	0.04–0.05	0.06–0.07	0.1–0.2

Table 2-12. Selected Properties of Polycarbonate

Properties	30% Glass	30% Graphite Fiber	40% Carbon Fiber
Relative density	1.41–1.43	1.33	1.36–1.38
Melting temperature (°C) T_m	—	—	—
T_g	150	149	—
Processing range (°F)	I:550–650	I:540–650	I:580–620
Molding pressure	10–30	—	15–20
Shrinkage	0.001–0.002	0.001–0.002	0.0005–0.001
Tensile strength	19,000–20,000	24,000	23,000–24,000
Compressive strength	18,000–20,000	26,000	22,000
Flexural strength	23,000	34,000–36,000	34,000–35,000
Izod impact (ft-lb/in.) (1/8 in. specimen)	1.7–2.0	1.8	1.5–2.0
Linear expansion (10^{-6} in./in./°C)	22–23	9	12.6
Hardness, Rockwell	M92, R119	R118	R119
Flammability	V-0/V-1	V-0	V-0
Water absorption, 24 h (%)	0.08–0.14	0.04–0.08	0.08–0.13

retainers, lighting lenses, appliance housings, and glazings. Coextruded packages are used for ovenproof frozen food trays and microwave pouches.

(Aromatic) Polyester

Aromatic polyester refers to a class of polymers containing rings of atoms. Wholly aromatic copolyesters may be based on terephthalic acid, p,p'-dihydroxybiphenyl, and p-hydroxybenzoic acid. A material based on the repeating p-oxybenzoyl units does not melt below its decomposition temperature. It must be sintered into bearing pads and seals. Compositions filled with metal powder may be flame or plasma spray coated on selected substrates, while other formulations may be melt processed on conventional equipment. Some formulations require processing temperatures over 800°F (427°C). As a class, these materials have outstanding strength, chemical resistance, weatherability, and thermal oxidative stability. With a decomposition temperature above 1000°F (538°C), the polymer forms a char layer that prevents burning and dripping.

Members of this class of materials are sometimes referred to as nematic, anisotropic, liquid crystal polymer (LCP), or self-reinforcing polymers. These terms try to describe the formation of tightly packed fibrous chains during the melt phase. It is the fibrouslike chain that gives the resin its self-reinforcing qualities. Properties may exceed those of fiber-reinforced conventional thermoplastics. Flexural moduli of 2.4×10^6 psi have been attained.

The unique microstructure of the polymers accounts for their extraordinary properties. The fibrous chains tend to arrange parallel to the mold flow, creating an anisotropy orientation effect. Mold designs must consider this morphology. No mold release or postcuring is generally required.

High-temperature applications include parts for engines, electrical connections, and aerospace and transportation components.

Polyetheretherketone (PEEK)

This aromatic (phenol-ether) structured thermoplastics has high temperature resistance and outstanding thermal properties. Service temperatures may exceed 450°F (232°C). The crystalline character contributes to its chemical and radiation resistance levels.

Polyetheretherketone may be processed on conventional thermoplastic equipment. Reinforced grades are molded into compressor parts, aerospace components, and electronic parts. Typical applications are for coatings of wire, cable, and other substrates.

Polyetherimide

This amorphous polymer is based on the repeating aromatic imide and ether units. Polyetherimide is heat resistant, strong, rigid, transparent amber, chemical resistant, and dimensionally stable (see Table 2-13). It may withstand continuous use temperatures in excess of 347°F (175°C). Reinforced and filled grades improve strength, temperature resistance, and creep. Carbon fiber- and mineral-filled grades are used for bearings and pads.

Table 2-13. Selected Properties of Polyetherimide

Properties	30% Glass Fiber Reinforced	30% Carbon Fiber
Relative density	1.49–1.51	1.39–1.42
Melting temperature (°C) T_m	—	—
T_g	215	215
Processing range (°F)	I:620–800	I:640–780
Molding pressure	10–20	10–20
Shrinkage	0.001–0.002	0.0005–0.002
Tensile strength	25,000–28,500	29,000–34,000
Compressive strength	23,500–24,000	32,000
Flexural strength	33,000–37,500	37,000–44,000
Izod impact (ft-lb/in.)		
(1/8 in. specimen)	1.7–2.0	1.2–1.6
Linear expansion		
· (10^{-6} in./in./°C)	20–21	10
Hardness, Rockwell	M125, R123	M127
Flammability	V-0	—
Water absorption, 24 h (%)	0.18–0.20	0.2

Polyetherimide may be processed on conventional thermoplastic equipment. All grades must be dried before processing.

Typical applications include radomes, electrical connectors, jet engine components, interior sheeting for aircraft, ovenproof cookware, flexible circuitry, composite structures, and food packaging.

Polyethersulfone (PES)

This amorphous structured thermoplastic is characterized by aromatic groups, which contribute to its high-temperature performance. Continuous service temperatures may exceed 356°F (180°C). It has good creep, chemical, radiation, and impact resistances. Transparent and reinforced or filled grades are also available.

The materials may be processed on conventional equipment.

Applications include projector lamp grills, sight glasses, sterilizable medical components, terminal blocks, printed circuit boards, radomes, and numerous aerospace composite components.

Polyethylene (PE)

The terms **polyolefin** and **ethenic polymers** are used to describe a family of polymers called polyethylene. The word **olefin** means oil forming and **ethenic** refers to ethylene materials. There are several processes used to link the ethylene molecules together (Ziegler, Phillips, and Standard Oil).

ASTM has divided polyethylenes into four groups:

Type	Density
I	(branched) 0.910–0.925 (low density)
II	0.926–0.940 (medium density)
III	0.941–0.959 (high density)
IV	(linear) 0.960 and above (high density to ultrahigh)

With increased density the properties of stiffness, softening point, tensile strength, crystallinity, and creep resistance increase. Increased density reduces impact strength, elongation, flexibility, and transparency. Glass fibers and fillers increase rigidity, tensile strength, and thermal resistance and lower creep.

The density range affects the structure, processability, and use. The molecular weight, molecular weight distribution, crystallinity, branching, copolymerization, and alloying also vary greatly (see Tables 2-14 and 2-15). (See High Molecular Weight–High-Density Polyethylene, Linear Low-Density Polyethylene, Ultrahigh Molecular Weight Polyethylene, and Very-Low-Density Polyethylene.)

Polyethylene may be cross-linked to convert a thermoplastic material to a thermoset.

Table 2-14. Selected Properties of Polyethylene Homopolymers

Properties	Branched	Linear
Relative density	0.917–0.932	0.918–0.935
Melting temperature (°C) T_m	106–115	122–124
$\qquad\qquad\qquad T_g$	—	—
Processing range (°F)	I:300–450	I:350–500
	E:250–450	E:450–600
Molding pressure	5–15	5–15
Shrinkage	0.015–0.050	—
Tensile strength	1,200–4,550	1,900–4,000
Izod impact (ft-lb/in.)		
(1/8 in. specimen)	No break	1.0–9.0
Linear expansion		
(10^{-6} in./in./°C)	100–220	—
Hardness, Rockwell	Shore D44–50	—
Flammability	HB	HB
Water absorption, 24 h (%)	<0.01	—

Table 2-15. Selected Properties of Polyethylene

Properties	Low and Medium Molecular Weight	High Molecular Weight	Ultrahigh Molecular Weight	30% Glass Fiber Reinforced
Relative density	0.939–0.960	0.947–0.944	0.94	1.18–1.28
Melting temperature ($°C$) T_m	125–132	125–135	125–135	120–140
T_g	—	—	—	—
Processing range ($°F$)	I:375–500 E:300–500	I:375–500 E:375–475	C:400–500	I:350–600
Molding pressure	5–20	—	5–20	10–20
Shrinkage	0.012–0.040	0.015–0.040	0.040	0.002–0.006
Tensile strength	3,000–6,500	2,500–4,300	5,600	9,000
Compressive strength	2,700–3,600	—	—	7,000
Flexural strength	—	—	—	11,000
Izod impact (ft-lb/ in.) (1/8 in. specimen)	0.35–6.0	3.2–4.5	No break	1.1–1.3
Linear expansion (10^{-6} in./in./ $°C$)	70–110	70–110	130	48
Hardness, Rockwell	—	—	R50	R75
Flammability	HB	—	—	V-0
Water absorption, 24 h (%)	<0.01	—	<0.01	0.02

Thermoplastic polyethylenes may be processed on all thermoplastic equipment. As the melt index goes down, the melt viscosity increases. The melt index and molecular weight distribution impact processing parameters. Generally, as molecular weight increases, melt index decreases. Low melt indexes are used for blow molding and injection molding of thick parts. (See Molding Processes.)

Polyethylenes as a family are characterized by excellent chemical resistance, high shrinkage after molding, flexibility, and high permeability to oxygen. Applications may include film packaging, bag liners, diaper liners, and household wrap. Blow molding examples include gasoline tanks, squeeze bottles, and food containers. They are injection molded into pails, dishpans, and luggage. Pipe and profile shapes are extruded, as are coated cables and wires.

High Molecular Weight–High-Density Polyethylene (HMW–HDPE)

This class of linear homopolymers or copolymers has densities greater than 0.941 and molecular weights (M_w) ranging from 250,000 to over 500,000. Butene, hexene, and octene are typical comonomers. The high molecular weight, density, and broad molecular weight distribution result in a tough, stress-crack resistant, stiff, abrasion- and chemical-resistant polymer. This also results in high melt viscosity. All grades must be stabilized with antioxidant and stabilizers. Most polymers are processed by extrusion processing. Principal applications include irrigation pipe, sewer liners, trash bags, and blow molded containers as large as 500 gal.

Linear Low-Density Polyethylene (LLDPE)

This polymer is produced by the copolymerization of ethylene and other alpha olefins. It is a linear rather than a branched molecule as in LDPE. Melt index, density, molecular weight distribution, and comonomer options make this a most versatile polymer with a wide range of properties. Low warpage, toughness, and excellent low-temperature impact strength are characteristic properties.

Linear low-density polyethylene may be processed on all thermoplastic equipment. Much of this material is made into food packaging. Ice bags, trash bags, industrial liners, garment bags, bottles, and extruded pipe are primary applications.

Ultrahigh Molecular Weight Polyethylene (UHMWPE)

The molecular weight of this very long, linear chained polymer ranges from 3×10^6 to 6×10^6. It is characterized as having a low coefficient of friction, toughness, good stress-crack resistance, and good impact strength at cryogenic temperatures.

It is normally manufactured into powder to be processed by special compression, sintering, extrusion, and injection techniques. Fillers and reinforcements may be added to enhance stiffness and reduce creep. Sintered and molded parts are sometimes postformed. Skived sheets are heated and formed between dies.

Typical applications include gears, bushings, packaging liners, coatings on wear surfaces, conveyors, sprockets, railcar and ship cargo liners, or other applications where a tough, abrasion-resistant, nonstick surface is needed.

Very-Low-Density Polyethylene (VLDPE)

This linear, nonpolar polyethylene is produced of materials similar to LLDPE. Densities range from 0.890 to 0.915. They may be produced with a wide range of properties depending on density, molecular weight, and molecular weight distribution. Typically, they are tough, flexible, chemical resistant, and have good thermal stability. They may be processed on all conventional thermoplastic equipment and must compete in application with products made from EVA, DDA, and LLDPE.

Tear-resistant films are made into disposable gloves and shrink packages and may be coextruded with other plastics to improve tear strength. Vacuum cleaner hoses, tubing, squeeze tubes, bottles, and container liners are additional applications.

Polyethylene Terephthalate (PET)

Polyethylene terephthalate is a melt condensation polymer from terephthalic acid or dimethyl terephthalate and ethylene glycol. To reduce crystallinity, PET may be copolymerized. Copolyesters of glycol-modified PET are called PETG. Clear shampoo and detergent bottles are familiar applications. PCTA copolyester is produced from cyclohexanedimethanol, terephthalic acid, and other dibasic acids. The result is an amorphous product made into clear film for packaging foods.

Polyethylene terephthalate polymers are saturated, linear polymers with high molecular weight. Polyester clothing fibers, tire cords, drafting films, and beverage bottles are the most familiar applications. These polymers offer high strength, stiffness, and dimensional stability in addition to chemical, heat, and creep resistance. Glass-reinforced and -filled grades offer improved mechanical, thermal, and chemical properties. (See Table 2-11.)

The polymer must be dried before processing using conventional processing equipment. Most container applications require PET to be biaxially oriented or stretched to improve creep resistance from carbonation pressure and acceptable gas barrier properties. Other applications for PET may include computer fans, impellers, casters for office chairs, cowl vent grills, and fuse holders.

Polyimide (Thermoplastic)

These aromatic polyimides are linear, thermoplastic materials with high temperature resistance, toughness, dimensional stability, and chemical and radiation resistances. They are obtained from condensation polymerization of an aromatic dianhydride and an aromatic diamine. Monomeric reactant, addition-type polyimide is used in composite structural components in aerospace applications.

Powders and solutions are used in processing. A sintering process is used to form laminated or compression molded parts. Most are cooled in the mold before demolding. Polyimide in a hot solution may be applied to coat houseware items or other substrates forming a finish similar to porcelain. Solutions may be used as the matrix in reinforcing cloth to produce a type of laminated vacuum-bag product. Autoclave processes use polyimide composites to form engine nozzle flaps. Fibers are made by continuous solution spinning and films by solution casting.

Applications include radomes, printed circuit boards, bearings, clothing, upholstery, wear strips, and dielectrics.

Polymethyl Methacrylate (See Acrylics)

Polymethylpentene

This polymer is an isotactically arranged aliphatic polyolefin of 4-methylpentene-1. Although the plastics is nearly 50% crystalline, it has a light transmission value of 90%. Transparency, low density, and relatively high melting point are distinguishing features.

Polymethylpentene may be processed on normal thermoplastic equipment. It is made into autoclavable medical equipment, syringes, light diffusers, lenses, food packaging, cook-in containers, coated paperboard for convenience foods, and components for microwave equipment.

Polyoxymethylene (See Acetals)

Polyphenylene Ether (PPE)

Polyphenylene polymers and alloys belong to a group of aromatic polyethers. Polyphenylene with no benzene-ring separation is very brittle, insoluble, and infusible. Polyphenylene ether copolymers are formed with an ether linkage. Phenol monomers, methyl groups, or alloys of polystyrene help provide a wide range of properties. Polyphenylene ether has a very high melt viscosity. Alloys such as PS help to lower melt viscosity and allow for processing on conventional equipment. The addition of fillers and reinforcements further increases impact strength, stiffness, thermal stability, and chemical resistance. Some glass fiber grades offer a modulus value over 1 × 10^6 psi.

These copolymers and alloys are used as appliance housings, plumbing fixtures, business machine cabinets, and electrical components.

Polyphenylene Oxide (PPO)

Another polyether polymer with similar properties to polyphenylene ether is polyphenylene oxide (poly 2,6-dimethyl-1, 4-phenylene oxide). To facilitate processing in the melt, commercial grades generally contain polystyrene. These polymers offer high heat deflection temperatures, flame retardancy, high impact strength, and excellent electrical properties (see Table 2-16).

They are processed on all conventional thermoplastic equipment and made into computer housings, video display terminals, pump impellers, radomes, fuse boxes, instrument panels, and business machine covers.

Table 2-16. Selected Properties of Polyphenylene Oxide

Properties	30% Glass Fiber Reinforced	40% Aluminum Flake
Relative density	1.27–1.36	1.45
Melting temperature (°C) T_m	—	—
T_g	100–110	—
Processing range (°F)	I:500–630 E:460–525	I:500–600
Molding pressure	12–40	10–20
Shrinkage	0.001–0.004	0.001
Tensile strength	16,000–18,500	6,500
Compressive strength	17,900	6,000
Flexural strength	20,000–23,000	9,500
Izod impact (ft-lb/in.) (1/8 in. specimen)	1.7–2.3	0.6
Linear expansion (10^{-6} in./in./°C)	14–25	11
Hardness, Rockwell	R115–116	R110
Flammability	HB	HB
Water absorption, 24 h (%)	0.06	0.03

Table 2-17. Selected Properties of Polyphenylene Sulfide

Properties	40% Glass Fiber Reinforced	Mineral and Glass Filled
Relative density	1.60–1.65	1.8–1.9
Melting temperature (°C) T_m	275–290	285–290
T_g	88	88
Processing range (°F)	I:600–675	I:600–650
Molding pressure	5–20	5–20
Shrinkage	0.002–0.004	0.004
Tensile strength	19,500–23,000	13,000–14,900
Compressive strength	2,100	11,000–23,000
Flexural strength	29,000–32,000	17,500–23,000
Izod impact (ft-lb/in.)		
(1/8 in. specimen)	1.4–1.5	0.5–1.0
Linear expansion		
(10^{-6} in./in./°C)	22	20
Hardness, Rockwell	R123	R121
Flammability	V-0	V-0
Water absorption, 24 h (%)	0.02–0.05	0.03

Polyphenylene Sulfide (PPS)

This rigid, crystalline polymer with benzene rings and sulfur links exhibits outstanding high-temperature stability and chemical and abrasion resistance (see Table 2-17). It is opaque and insoluble in most chemicals. The polymer wets fibers and fillers well. Numerous reinforced grades are available that significantly increase mechanical properties.

These polymers may be processed on conventional thermoplastic equipment. Applications include small appliances, range components, hair dryers, submersible pump enclosures, and computer components.

Polypropylene (PP)

Most of the properties of this polymer are related to melt index, molecular weight, degree of crystallinity, molecular weight distribution, and proportion of isotactic to atactic structure. Polypropylene is made from propylene gas and a methyl group. It is not surprising that PP and PE have many similar properties. Polypropylene's key properties are low density (0.90–0.91), high heat resistance, stiff-

Table 2-18. Selected Properties of Polypropylene

Properties	40% Glass Fiber	30% PAN Carbon Fiber
Relative density	1.22–1.23	1.04
Melting temperature (°C) T_m	168	168
T_g	—	—
Processing range (°F)	I:450–550	I:360–470
Molding pressure	10–25	—
Shrinkage	0.003–0.005	0.001–0.003
Tensile strength	8,400–15,000	6,800
Compressive strength	8,900–9,800	—
Flexural strength	10,500–22,000	9,000
Izod impact (ft-lb/in.)		
(1/8 in. specimen)	1.4–2.0	1.1
Linear expansion		
(10⁻⁶ in./in./°C)	27–32	—
Hardness, Rockwell	R102–111	—
Flammability	HB	HB
Water absorption, 24 h (%)	0.05–0.06	0.12

ness, and chemical resistance (see Table 2-18). Ethylene or other elastomers may be alloyed to improve impact properties. Selected mechanical and thermal properties are improved with the addition of fillers and reinforcements.

Polypropylene is processed on all thermoplastic equipment. It is used extensively as fiber and filament in carpeting, upholstery fabric, and bristles. Other applications include wrap for snack foods, tobacco products, and cheese, orange juice cups, margarine tubs, storage battery cases, luggage, syringes, and toys. (See Figure 2-6.)

Polystyrene (PS)

Styrene is chemically known as vinyl benzene. This monomer is obtained from ethyl benzene and polymerized by bulk, solvent, emulsion, or suspension techniques. Polystyrene accounts for about 20% of all the thermoplastics in commercial use. It is a low-cost, atactic, amorphous thermoplastic. There are three general grades or classes of polystyrene. The general purpose grade is sometimes called crystal (unfortunately). This implies that it is an amorphous (not crystalline) glass-clear polymer. This grade has high stiffness and low impact resistance. These should not be considered weather resistant

Figure 2-6. Multilayer bottles are coextrusion blow molded with inner and outer layers of PP enclosing an oxygen barrier layer of ethylene vinyl alcohol (EVOH). (Continental Can Company, Continental Plastic Containers Division)

unless suitably modified (see Table 2-19). Rubber-modified classes are called impact graded. Formulations of up to 10% rubber improve impact resistance but reduce stiffness and transparency. Expandable grades come in the form of gas-filled beads. As the petroleum ether and beads are heated, a cellular part is produced in the mold cavity as the gas causes the bead to swell or expand. (See Expanded, Foamed.) These grades are used for making low-density (foamed) expanded material for packing and heat insulating purposes.

Crystal and impact grades are processed on all conventional thermoplastic equipment.

General applications for all grades include window moldings, foamed sheets for meat trays and egg cartons, cassettes, reels, decorative display boxes, wall tile, bottle caps, and vacuum-formed containers of all kinds.

Polysulfone

Polysulfone is a copolymer of 4,4-dichlorodiphenylsulfone and bisphenol A. It is selected when good thermal stability, transparency, and rigidity at high temperatures are required. These properties make polysulfone a replacement material for markets using glass

Table 2-19. Selected Properties of Polystyrene

Properties	High Heat-Resistant Styrene Copolymer	Styrene Copolymer/ Polycarbonate	20% Long and Short Glass	Styrene/ Acrylonitrile 20% Glass
Relative density	1.05–1.08	1.13–1.15	1.20	1.22–1.40
Melting temperature (°C) T_m	—	—	—	—
T_g	—	—	115	120
Processing range (°F)	I:425–540 E:400–500	I:530–580	I:400–550	400–500
Molding pressure	—	1–1.5	10–20	10–20
Shrinkage	0.002–0.006	0.007	0.001–0.003	0.001–0.003
Tensile strength	4,600–5,800	6,800–8,000	10,000–12,000	15,500–18,000
Compressive strength	—	—	16,000–17,000	17,000–21,000
Flexural strength	8,500–10,500	8,000–13,000	14,000–18,000	20,000–22,700
Izod impact (ft-lb/in.) (1/8 in. specimen)	1.5–4.0	10–12	0.9–2.5	1.0–3.0
Linear expansion (10^{-6} in./in./°C)	67	—	39.6–40	23.4–41.4
Hardness, Rockwell	L75–95	—	M80–95, R119	M89–100, R122
Flammability	HB	—	V-0	
Water absorption, 24 h (%)	0.1	0.20	0.07–0.01	

and metal. Continuous service temperatures may exceed 374°F (190°C) (see Table 2-20).

It comes in several grades and may be reinforced and filled. All grades must be dried before molding with conventional thermoplastic processing equipment. The light amber color of the plastics is a result of the addition of methyl chloride, which ends polymerization.

Applications include substrate for circuit boards, microwave oven cookware, coffee makers, camera cases, and as a matrix for carbon-fiber composites for aircraft parts.

Table 2-20. Selected Properties of Polysulfone

Properties	30% Glass	30% Carbon Fiber	Polyarylsulfone	Polyether-sulfone 30% Carbon Fiber
Relative density	1.46	1.36	1.29–1.37	1.47–1.48
Melting temperature ($^\circ$C) T_m	—	—	—	—
T_g	189–190	—	220	—
Processing range ($^\circ$F)	I:600–700	I:600–700	I:630–800 E:620–750	I:680–720
Molding pressure	—	10–20	5–20	10–20
Shrinkage	0.001–0.003	0.001	0.007–0.008	0.005–0.002
Tensile strength	14,500	23,500	9,000	26,000–30,000
Compressive strength	19,000	25,000	—	22,000
Flexural strength	20,000	32,000	12,400–16,100	36,000–38,000
Izod impact (ft-lb/in.) (1/8 in. specimen)	1.1	1.2	1.6–12	1.2–1.6
Linear expansion (10^{-6} in./in./$^\circ$C)	25	6	31–49	10
Hardness, Rockwell	M90–100	M80	—	R123
Flammability	V-0	V-0	—	
Water absorption, 24 h (%)	0.3	0.15	—	0.29

Polyvinyl Acetate (PVAc)

Polyvinyl acetate is produced from reactions of acetic acid and acetylene. It is a relatively soft material used primarily for coatings and adhesives. PVAc is usually in an emulsion form such as white glues.

Polyvinyl Alcohol (PVAl)

Polyvinyl alcohol is water soluble and is used as a coating and an adhesive. It has good film-forming ability and resistance to organic solvents.

Polyvinyl Butyral (PVB)

Polyvinyl butyral is produced from polyvinyl alcohol and used as an interlayer in (laminated) safety glass.

Polyvinyl Chloride (PVC)

Through common usage, the vinyl plastics are those polymers with the vinyl name. Many authorities limit their discussion to include only polyvinyl chloride and polyvinyl acetate. The family of vinyls comprises polyvinyl chloride, polyvinyl acetate, polyvinyl alcohol, polyvinyl butyral, polyvinyl acetal, polyvinylidene chloride, polyvinyl fluoride, and others.

In terms of tonnage, polyvinyl chloride ranks second after polyethylene as the leading plastics produced. PVC is made by reacting acetylene gas with hydrochloric acid. The monomer has one chloride atom substituted for a hydrogen atom. The four basic polymerization processes in diminishing order are suspension, bulk, emulsion/dispersion, and solution. Characteristics of PVC are chemical, corrosion, and weather resistance. PVC is most often highly plasticized or used as dispersions. Homopolymers generally contain about 56% chlorine with the exception of chlorinated PVC with about 67% chlorine. Added chlorine increases heat deflection temperatures. Chlorinated PVC may be used for hot water piping in homes. A wide variety of properties (see Table 2-21) is available with the addition of monomers of ethylene, vinyl acetate, styrene, vinylidene, acrylic acid, or alloys of PE, ABS, EVA, and EPDM.

Table 2-21. Selected Properties of PVC

Properties	Barrier Plasticized	PVC 15% Glass
Molding qualities	Good	Good
Relative density	1.68–1.72	1.54
Melting temperature (°C) T_m	172	—
$\quad T_g$	−15	75–105
Processing range (°F)	E:340–400	E:270–405
Molding pressure	5–30	8–25
Shrinkage	0.005–0.025	0.0001
Tensile strength	3,500	9,500
Compressive strength	2,000–2,700	9,000
Flexural strength	—	13,500
Izod impact (ft-lb/in.) (1/8 in. specimen)	0.3–1.0	1.0
Linear expansion (10^{-6} in./in./°C)	190	—
Hardness, Rockwell	R98–106	R118
Flammability	V-0	V-0
Effect of sunlight	Stabilizer	Stabilizer
Effects of solvents	Ketones	Ketones
Water absorption, 24 h (%)	0.1	0.01

Because PVC is thermally sensitive, heat stabilizers, lubricants, and plasticizers are commonly used to aid in processing the melt. Over half of the PVC is made into rigid products such as pipes, flooring, window sashes, gutters, and exterior siding. Most thermoprocessing techniques are used with added protection from corrosive and toxic halogens. Plastisols and organosols are made into cast parts or used to coat a variety of substrates. Everyone is familiar with plasticized, flexible PVC, which can be used as meat wrap, wall coverings, upholstery, footwear, and furniture.

Polyvinyl Fluoride (PVF)

Polyvinyl fluoride resembles PVC in chemical properties but its mechanical properties are superior because the fluorine atom's size permits molecules to pack in a crystalline fashion. It is an exceptionally good outdoor product. PVF is used for film and coating on many substrates.

Polyvinyl Formal

Polyvinyl formal is produced from polyvinyl acetate and formaldehyde. It is used as a coating for metal containers and as wire enamels.

Polyvinylidene Chloride (PVDC)

A common name for PVDC has come from a well-known food wrapping film trade name, Saran. This substance is similar to polyvinyl chloride but it contains one more chlorine atom. Polyvinylidene chloride is commonly used as a copolymer with polyvinyl chloride, acrylonitrile, or acrylate esters. PVDC is known for low gas and liquid permeability. Films are multiaxially oriented to increase strength. Latex or solution coatings are used on paper and other substrates as protective and barrier layers. Reinforced and alloyed melt processible grades are available. Care should be taken to protect metal surfaces because of the corrosive effects of chlorine. Adequate venting should be provided to protect the operator. Most thermoplastic processing is used.

Typical applications include monolayer and multilayer barrier films, sheets, and coatings. Monolayer films are widely used in the home. Multilayered films are used to package food, contain trash, line pipes, and package pharmaceuticals and cosmetics.

Polyvinylidene Fluoride (PVDF)

This polymer has properties similar to those of PVF. It may be melt processed and made into films, laminates, and wire coverings.

(Olefin-Modified) Styrene–Acrylonitrile (OSA)

By tailoring the molecular weight and monomer ratios of saturated olefinic elastomer with styrene and acrylonitrile, a tough, heat- and weather-resistant polymer is produced.

Although the material is slightly hygroscopic, it may be processed by conventional thermoplastic techniques.

Much of the polymer is coextruded over other substrates. This exterior layer may be used for topper covers, boat hulls, animal shelters, automobile trim, and decorative wood and metal construction panels.

Styrene–Acrylonitrile (SAN)

This thermoplastic material is a copolymer of styrene and acrylonitrile. It is produced by emulsion, suspension, or continuous bulk polymerization. The addition of acrylonitrile content will enhance physical properties. This amorphous copolymer has good tensile and flexural strength. Chemical resistance, toughness, modulus, and tensile strength are improved by biaxial orientation in processing.

The copolymer is slightly hygroscopic but may be processed on conventional equipment.

Applications include blender bowls, instrument lenses, syringes, cosmetic containers, pen and pencil barrels, knobs, and refrigerator compartments.

Styrene–Butadiene Plastics (SBP)

Styrene butadiene block copolymer consists of two blocks of styrene repeating units separated by a block of butadiene. This is in contrast to styrene butadiene rubber (SBR) which is a thermosetting, cross-linked elastomer. This does not imply that some cross-linking may occur. This amorphous copolymer is recognized for its transparency, high impact strength, and high gloss.

It may be processed on conventional thermoplastic equipment and may be reinforced, filled, or alloyed with other polymers such as PS, SAN, PP, or PC.

Applications include food trays, bottles, jars, blister packs, overwrap, tool handles, and integral hinge containers.

Styrene–Maleic Anhydride (SMA)

This polymer is obtained by the free radical copolymerization of a mixture of maleic anhydride and styrene. For impact grades, butadiene monomers may be used with the SMA copolymer. Other alloys such as ABS may be combined for improved impact strength. Reinforcements and fillers generally improve heat, creep, and chemical resistances. Mechanical properties and tensile strength may be doubled with reinforced grades.

In general, the SMA copolymers and terpolymers have similar properties to ABS grades.

These materials may be processed by most conventional processing techniques. Applications include vacuum cleaner housings, mirror housings, thermoformed headliners, fan blades, and electrical connectors.

Thermosetting Polymers

The following discussion is to acquaint the reader with a brief overview of thermosetting materials. Although selected grades and data are mentioned, it is not possible to list all the possible combinations or the improvements or changes made in properties by the addition of other constituents. The data given with each polymer family should be used only for comparison purposes. New, more current data are available from several sources. (See Thermoplastic Polymers.)

You should recall that thermosetting materials undergo a chemical reaction and become "set." It is the low molecular weight, polyfunctional resin matrix that determines if a material will be thermoplastic or thermosetting. The thermosetting resin (matrix) component usually contains monomers, curing agents, hardeners, inhibitors, and plasticizers. The other major component used with the resin matrix may consist of mineral or organic particles, inorganic or organic fibers, chopped or woven cloth, paper, or other laminars.

Once the matrix has undergone a cure and become set, reheating will not soften these materials. In most thermosetting materials, polymer chains are cross-linked by polymerization under heat and pressure or by a catalyzing operation. It is the strong covalent bonds between chains that provide greater thermal stability and creep resistance than is possible in most thermoplastics. The weaker secondary bonds in thermoplastic materials allow most of them to be melt processed. Thermosetting materials are generally more difficult to process and require longer molding cycles. (See Molding Processes.)

In general, because most thermosetting matrices are high molecular weight materials, they are inherently brittle or fractured. Most are filled or reinforced to improve thermal stability, deformation resistance, dimensional stability, and hardness, to meet design and processing configurations, or to lower costs. (See Design Practice.) Thermosetting materials are supplied in a number of available forms. Liquid **resins** usually consist of a partially polymerized monomer or monomers. This may be considered an A-stage material and soluble in certain liquids. Any one or a combination of energy sources such as heat, light, or electron energy may initiate a polymerization of the monomers. Catalysts, promoters and other additives are used to polymerize the syrupy resins at room temperature. The heat created by the chemical reaction is often called the exotherm. It is this exotherm that cures (sets) the material. External forms of radiation including thermal, wave, or particle energy may be used to aid or complete the curing reaction. Resins are preferred in many processing applications because they have an excellent wetting (good adhesion to reinforcements) effect in bonding laminar, fibrous, or particle reinforcements. Many resins are compounded and used in casting operations. (See Casting Processes.)

The remaining forms are compounded and supplied as powders, granules, preforms, bulk molding compounds, or sheet molding compounds. These available forms are then preformed, that is, caused to take the shape of the mold, and polymerized to form the finished product. **Powders** and **resins** may be used in coating substrates. **Granules** are often made from B-stage resins and other additives. During this intermediate state, resins can be heated and caused to flow to the desired shape. The term C-stage refers to the cured insoluble and infusible polymer. It is during the B-stage that most thermosetting materials are molded into the C-stage products. Powders and granules may be pressed into **preform** shapes. Preforms are often referred to as pills, tablets, biscuits, or premolds. Carefully measured mixtures shaped into preforms help speed the molding cycle. Cycle times are also shortened by preheating the premix. **Bulk molding compounds (BMC)** are sometimes called the premix, gunk, putty, dough, or slurry. They are usually compounded from monomer resin, reinforcements, fillers, and other additives. The thick puttylike materials are commonly formed into rope, slugs, or other shapes to aid molding and handling operations. **Prepreg** refers to compounds ready to mold, generally in sheet or tape form. The prepreg is composed of cloth, mat, or paper impregnated B-stage resin. (See Reinforcements, BMC, SMC, TMC, or XMC.)

Sheet molding compounds (SMC) are sometimes called the flow mat or mold mat. They are similar to bulk molding compound preforms or premixes. SMC are precombined sheets or resin, reinforcement, fillers, fabrics, and/or other additives with an upper and lower carrier film. (See Figure 2-7.) In this form, the sheet is easily handled and adapted to mass production techniques. This film is removed prior to molding. Thermosetting plastics also come as **adhesives** in powder, resin, film, or dispersion forms. **Cellular** or **foamed** forms are available but most are molded in situ (in place). Only a few are available in **fiber** form.

Distinguishing characteristics of individual groups of thermosetting materials are arranged in alphabetical order. Only commercially significant polymers are included.

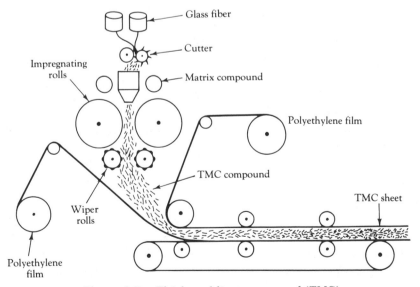

Figure 2-7. Thick molding compound (TMC).

Alkyd

The word **alkyd** is derived from the *al* in alcohol and the *cid* in acid. It is usually pronounced al'-kid. Alkyd molding compounds are composed of unsaturated polyesters cross-linked with a diallyl phthalate monomer, filler, and other additives. They are unsaturated polyesters cross-linked with other monomers. It is possible to cross-link unsaturated polyester but the reaction is slow. Styrene is the most widely used cross-linking monomer in other polyesters. Some suppliers term alkyd as compounds with the lowest amount of mon-omer used to perform the cross-link. (See Polyester.) Alkyd, MCB, and preforms are molded by all thermosetting techniques. Baked alkyd finishes on metal and oil-based alkyd house paints are still used today.

Allyl

Allyl esters (polyallyl ester) are derivatives of allyl alcohol or the allyl esters of dibasic acids. The three most important allylic mon-omers and prepolymers are diallyl phthalate (DAP), diallyl isophthal-ate (DAIP), and allyl diglycol carbonate.

These monomers are used to make molding compounds and preimpregnated glass cloths and papers. Many are used as cross-linking agents for other monomer chains. (See Polyester, Alkyds.) Because of the difunctionality of the acid involved, these materials are all thermosetting resins when fully cured. They are radiation-curable, light, dimensionally stable, heat abrasion- and chemical-resistant materials (see Table 2-22). Allyl diglycol carbonate has been used for light plastics eye lenses.

Allylic materials are processed by all modern thermoset molding techniques. Filled and fiber-reinforced grades increase mechanical, electrical, and thermal resistances. DAP is molded into insulators, circuit boards, switches, and electrical boxes. Preimpregnated cloth, mat, or other laminar materials are made into decorative laminates, aircraft panels, radomes, missile parts, and ducting. (See Figure 2-8.)

Amino

Amino polymers are formed by the interaction of amines or amides with aldehydes. It is of some significance that a major component of these materials is not based on fossil fuels but on the element nitrogen. There are many polymers in this family that have been

Table 2-22. Selected Properties of Allyl

Properties	Glass Filled
Relative density	1.70–1.98
Melting temperature (°C) T_m	Thermoset
T_g	—
Processing range (°F)	C:290–280
	I:300–350
Molding pressure	2,000–6,000
Shrinkage	0.0005–0.005
Tensile strength	6,000–11,000
Compressive strength	25,000–35,000
Flexural strength	9,000–20,000
Izod impact (ft-lb/in.) (1/8 in. specimen)	0.4–15.0
Linear expansion (10^{-6} in./in./°C)	10–36
Hardness, Rockwell	E80–87
Flammability	V-0
Water absorption, 24 h (%)	0.12–0.35

Figure 2-8. Epoxy, melamine, polyester, and phenolics are used as the matrix in these composite products. (Spaulding Fibre Company)

developed and used for special applications. Only two are currently of appreciable commercial importance—urea–formaldehyde and malamine–formaldehyde. Urea and melamine are readily combined with formaldehyde as addition monomer products. These monomers are then used as coatings and adhesives, or filled and made into various molding compounds. Both monomers are used as cross-linking agents with other monomer chains. A number of curing agents, catalysts, plasticizers, or other additives are normally compounded into the monomeric resin base before molding. The shelf life of these products is limited. To speed cure rates and lower cycle times, heat or other forms of radiation are commonly used. During final C-stage polymerization, a condensation reaction occurs, producing water.

Melamine–Formaldehyde (MF)

Melamine–formaldehyde (MF) is easily molded, nontoxic, easily colored, and scratch, heat, and stain resistant. Urea and melamine plastics are frequently grouped together as one material because of their similar chemical structures, properties, and applications (see Table 2-23). Melamine products are harder and more water resistant. Melamine resins are more expensive but combine easily with a greater variety of fillers. The largest single use of melamine–formaldehyde

Table 2-23. Selected Properties of Melamine Formaldehyde

Properties	Glass Fiber Reinforced
Relative density	1.5–2.0
Melting temperature (°C) T_m	Thermoset
T_g	—
Processing range (°F)	C:280–350
Molding pressure	2–8
Shrinkage	0.001–0.006
Tensile strength	5,000–10,500
Compressive strength	20,000–35,000
Flexural strength	14,000–23,000
Izod impact (ft-lb/in.) (1/8 in. specimen)	0.6–18
Linear expansion (10^{-6} in./in./°C)	15–28
Hardness, Rockwell	M115
Flammability	V-0
Water absorption, 24 h (%)	0.09–1.3

is in the manufacture of tableware. Alpha cellulose or other additives are used to form shaver housings, mixing bowls, appliance housings, and knobs.

Urea–Formaldehyde (UF)

Urea–formaldehyde (UF) is used extensively as adhesives in the manufacture of furniture, plywood, and chipboard. A matrix of about 10% resin is pressed together with wood chips to produce chipboard. Chipboard, or plywood bonded with this resin, is suitable only for interior use. Molding compounds with latent catalysts and other additives are molded into appliance knobs, handles, toaster bases, circuit breakers, and toilet seats. In the water-soluble formalin stage, they are used as bonding and sizing agents in the fabrication of paperboard, corrugated cardboard, and bags. At one time monomeric surface coatings were applied on many appliances or machine equipment surfaces followed by a baking or cure cycle. Other durable, faster curing resins have replaced some of these applications. Uncured, partially polymerized plastics and foams degas and this may result in allergy and flulike symptoms in some people.

Epoxy (EP)

The epoxy group is also called the epoxide, oxirane, or ethoxyline group. The family gets its name from the epoxide reactive group in which an oxygen atom is joined to each of two carbon atoms already united in some other way. The major characteristics of epoxies as a family include excellent adhesion, mechanical properties, low moisture absorption, chemical resistance, little shrinkage, and ease of processing (see Table 2-24). Epoxy resins are among the best matrix materials for many composites. There are a number of resin formulations and a variety of curing agents. Through a combination of different processing techniques, a variety of resin formulations, curing agents, and other additives, there is almost unlimited property versatility.

This resin may be used and cross-linked with phenolics, melamines, polyamides, ureas, polyesters, and some elastomers. Epoxy resins are cross-linked by a number of agents such as amines, anhydrides, and acids. Most formulations result in a three-dimensional, cross-linked thermosetting matrix. There are linear, high

Table 2-24. Selected Properties of Epoxy

Properties	Bisphenol Glass Fiber Reinforced	Novolak Mineral and Glass Filled	Aluminum Filled
Relative density	1.6–2.0	1.85–1.94	1.4–1.8
Melting temperature			Thermoset
($°$C) T_m	Thermoset	Thermoset	—
T_g	—	—	
Processing range	C:300–330	T:340–400	
($°$F)	T:280–380		
Molding pressure	1–5	0.5–2.5	
Shrinkage	0.001–0.008	0.004–0.007	0.001–0.005
Tensile strength	5,000–20,000	6,000–15,500	7,000–12,000
Compressive strength	18,000–40,000	30,000–48,000	25,000–33,000
Flexural strength	8,000–30,000	10,000–21,000	8,500–24,000
Izod impact (ft-lb/in.) (1/8 in. specimen)	0.3–10.0	0.4–0.45	0.4–1.6
Linear expansion (10^{-6} in./in./$°$C)	11–50	35	5.5
Hardness, Rockwell	M100–112		M55–85
Flammability	V-0	V-0	V-0
Water absorption, 24 h (%)	0.04–0.20	0.17	0.1–4.0

molecular weight thermoplastic formulations. Most resins are low molecular weight until the curing agent is added and the matrix becomes fully cross-linked into one large molecule.

More than 90% of the epoxy resins are prepared from diglycidyl ether of bisphenol A (DGEBA). The general formula of epoxy resins based on bisphenol A and epichlorhydrin represents diglycidyl ethers since they contain two glycidyl ether groups. In some literature, these polymers are called polyethers.

Common curing agents in the amine group are diethylenetriamine (DETA) and triethylenetetramine (TETA). A number of poly-functional amines are used as curing agents. Most will proceed at room temperature. Aliphatic amines cause high exothermic reaction with the monomer. Aromatic amines require elevated heat to cure. Cyclic acid anhydrides are widely employed as curing agents. Nadic methyl anhydride (NMA) and malic anhydride (MA) are commonly used, but usually require careful storage to prevent degradation. In most curing applications, heat is required to initiate full cure. Boron trifluoride (Lewis acid) curing agents are popular because they cause very rapid (in minutes) polymerization of the epoxy resin.

It is beyond the scope of this text to describe many of the modified epoxy formulations. As a rule, all classes in the family cure with little molecular reorientation and no by-products. The result is a tough, relatively strain-free matrix with little shrinkage (less than 4%). Low shrinkage causes epoxies to adhere well to other substrates. Coupling agents are also used to improve this adhesion. (See Reinforcements.)

Some epoxies are compounded with fillers, reinforcements, and latent catalysts and molded into terminal posts, switch gears, insulators, bushings, brackets, and other structural composites. Some are used as castings for potting and encapsulation of electronic components. Excellent adhesion, chemical resistance, and toughness make epoxies a major choice for coating concrete, appliances, pipes, tanks, ship bulkheads, and hulls. The adhesive and wetting qualities of epoxies are used to adhere most substrates.

Epoxy compounds are also used in tooling for dies and molds. Particulate or fibrous reinforcing reduces cost and improves dimensional stability of these economical toolings. Epoxies are most important as composites. The matrix is used in laminar, fibrous, and particulate composite materials and structures. Laminated and reinforced structures are used to produce radomes, circuit boards, and aircraft and aerospace components. (See Figure 2-9.) Filament wound composite structures provide high strength-to-weight ratios for rocket motor housings, pressure vessels, pipe, and tanks. (See Figure 2-10.)

Phenolics

Phenolics are known chemically as phenol–formaldehyde (PF). Phenol–formaldehyde polymers were the first wholly synthetic material used. The plastics is a hard, dark, crystalline product. As a class of materials they have been called the workhorse of the thermosetting plastics. Although they have been somewhat eclipsed by newer polymers, their low cost and wide range of compounding formulations will keep them an important material. Although the monomer solution of phenol is commercially used, cresols, xylenols, or resorcinols may be used. Furfural may replace the formaldehyde.

As a family, phenolics have excellent dimensional stability and

Figure 2-9. Graphite/epoxy composite is used on the tail and horizontal stabilizer of the F-20 Tigershark tactical fighter. These workers are laying out the graphite reinforcement layers. The epoxy matrix is then cured under high temperature and pressure. (Northrop Corporation)

Figure 2-10. Glass/epoxy filament wound rocket motor case. (Structural Composite Industries)

74

heat, thermal, chemical, and electrical resistances (see Table 2-25). They are available in a number of forms including flake, powder, liquid, or compounds. Compounds are formulated with fillers and reinforcements to lower cost and extend properties. There are two main PF types: one-stage and two-stage resins. One-stage resins are called resols. **Resols** are soluble, low in molecular weight, and can form large molecules without additional curing agents during the molding cycle. A cross-linking reaction occurs during the hot molding cycle and results in an infusible, thermosetting material. Two-stage resins are linear and thermoplastic unless compounds capable of forming cross-linkage are added. They are called two-stage resins because some agent must be added before molding. The low molecular weight two-stage resins are also called **novolacs**. The soluble thermoplastic novolac is referred to as an A-stage resin. The B-stage novolac is compounded with fillers, fibers, lubricants, and hexamethylenetetramine. It is sold as prepregs, granules, or powders. Heat and pressure are used to activate the polymerization process and turn the B-stage compound into an insoluble, infusible, C-stage material.

Phenolics are used as adhesives or matrix binders in plywood, waferboard, grinding wheels, brake linings, clutch plates, and wood

Table 2-25. Selected Properties of Phenol–Formaldehyde

Properties	High-Strength Glass Fiber Reinforced	Asbestos Filled
Relative density	1.69–2.0	1.45–2.0
Melting temperature (°C) T_m	Thermoset	Thermoset
T_g	—	—
Processing range (°F)	C:300–380	C:290–400
	I:330–390	I:330–900
Molding pressure	1–20	2–20
Shrinkage	0.001–0.004	0.001–0.009
Tensile strength	7,000–18,000	4,500–7,500
Compressive strength	16,000–70,000	20,000–35,000
Flexural strength	12,000–60,000	7,000–14,000
Izod impact (ft-lb/in.)		
(1/8 in. specimen)	0.5–18.0	0.26–3.5
Linear expansion		
(10^{-6} in./in./°C)	8–21	10–40
Hardness, Rockwell	E54–101	M105–115
Flammability	V-0	V-0
Water absorption, 24 h (%)	0.03–1.2	0.1–0.5

particleboard. Impregnated fabric, wood, paper, or other plastics are used to make electrical circuit boards, gears, bearings, and composite structures. The paper core stock for decorative laminates is made of impregnated phenolics and decorative malamine–formaldehyde surface. Phenolics are dark opaque materials and may not be used as decorative cover stock. Filled molding compounds usually contain about 40% resins, 45% filler, and other additives. They are molded by conventional thermosetting practices into fuse blocks, handles, knobs, circuit breakers, light sockets, and appliance housings. (See Amino.)

Polyester

The word **polyester** is derived from two chemical processing terms, **poly**merization and **ester**ification. Unsaturated esters have carbon-to-carbon bonding sites that are reactive; that is, they are ready to attach to other atoms or molecular chains to form cross-linkage. Saturated materials lack these bonds.

There are a number of polyesters available: linear unsaturated polyesters, linear saturated polyesters of low molecular weight, linear saturated polyesters of high molecular weight, network polyesters, polyallyl esters, and polycarbonates.

Linear unsaturated polyesters may be produced from glycols, dibasic acids and other monomers. Propylene glycol and maleic acid (or maleic anhydride) are common components. Styrene is the most widely used cross-linking monomer. A common accelerator added to the resin is cobalt. Initiating room-temperature cure are methyl ethyl ketone peroxide and cyclohexanone peroxide. Benzoyl peroxide may be used at elevated-temperature curing. Linear unsaturated polyesters are commonly used in conjunction with glass fiber to produce composite structures such as automobile bodies, shower stalls, building panels, and boat hulls. (See Figure 2-11.) Casting grades are available for water clear embedments, electronic encapsulation, or cultured marble vanities and sinks. Filled and reinforced molding compounds may be molded into switch boxes and breaker components. (See Pultrusion, Casting.)

Linear saturated polyesters of low molecular weight (500–8000) are used mainly as plasticizers in other polymers.

Linear saturated polyesters of high molecular weight (>10,000) are formed by the condensation of ethylene glycol and terephthalic

Figure 2-11. Nine foot diameter pipe was made from a matrix of polyester. Each of these glass/polyester pipe sections are 50 feet long. (ICI Americas, Incorporated)

acid. This polymer is a thermoplastic material. (See Polyethylene Terephthalate.)

Network polyesters are derived from phthalic anhydride and glycerol. These polymers are called alkyds. (See Alkyds.) They are used as thermosetting molding compounds and coatings.

Polyallyl esters are produced from allyl alcohols. The two most important members are diallyl phthalate (DAP) and diallyl isophthalate (DAIP). Both are thermosetting when fully cured. (See Allylics, Allyl.)

Polycarbonates may be considered as polyesters of carbonic acid and polyhydroxy compounds. These aromatic materials are thermoplastic. (See Table 2-26) (See Polycarbonate.)

Polyimide (Thermoset)

Thermoplastic and thermosetting polyimides are produced. (See Polyimide, Thermoplastic.) The aromatic diamine resin may be converted to a cross-linked, infusible mass. Films are usually cast from

Table 2-26. Selected Properties of Polyester

Properties	Preformed, Chopped Roving	Woven Cloth	SMC
Relative density	1.35–2.30	1.50–2.10	1.65–2.6
Melting temperature ($^\circ$C) T_m	Thermoset	Thermoset	Thermoset
T_g	—	—	—
Processing range ($^\circ$F)	C:170–320	C:73–250	C:270–350
Molding pressure	0.25–2	0.3	0.3–1.2
Shrinkage	0.0002–0.002	0.0002–0.002	0.00 ± −0.004
Tensile strength	15,000–30,000	30,000–50,000	8,000–25,000
Compressive strength	15,000–30,000	25,000–50,000	15,000–30,000
Flexural strength	10,000–40,000	40,000–80,000	10,000–36,000
Izod impact (ft-lb/in.) (1/8 in. specimen)	2–20	5–30	7–22
Linear expansion (10^{-6} in./in./$^\circ$C)	20–50	15–30	20
Flammability	HB	HB	V-0
Water absorption, 24 h (%)	0.01–1.0	0.05–0.5	0.1–0.25

the poly(amic acid) solutions. Molded polyimide parts are produced by compressing and sintering operations. They are made into aircraft engine parts, bushings, bearings, IC chip carriers, gears, and thermal insulation for wires. Films are used in dielectrics and flat cable devices. Glass fiber-reinforced grades have a flexural modulus exceeding 3×10^6 psi. Thermosetting polyimides have excellent creep, chemical, thermal, and wear resistances.

Polyurethane (PU)

The term **polyurethane** refers to the reaction of polyisocyanates and polyhydroxyl groups. These polymers have only limited use as solid plastics. They are best known as coatings, elastomers (PUR), and foams (see Table 2-27).

There are a number of formulations of foam. Most rigid, low-density foams are made from polymeric isocyanates (PMDI) and made into thermal insulation for piping tanks and architectural applications. High-density grades are used as wood replacement parts on door and picture frames, cabinet drawer fronts, and electronic cab-

Table 2-27. Selected Properties of Polyurethane

Properties	50–65% Mineral Filled	10–20% Glass Fiber	30% PAN Carbon Fiber
Relative density	1.16–2.1	1.22–1.36	1.33
Melting temperature			
$(°C)\ T_m$	Thermoset	—	—
T_g	—	120–160	—
Processing range (°F)	25 (casting)	I:360–410	I:410–450
Molding pressure	—	8–11	—
Shrinkage	0.5–0.7	0.007–0.010	0.001–0.002
Tensile strength	1,000–5,360	4,800–6,500	13,000
Compression strength	—	5,000	—
Flexural strength	—	5,500–6,200	9,000
Izod impact (ft-lb/in.)			
(1/8 in. specimen)	—	14–No break	10
Linear expansion			
$(10^{-6}\ in./in./°C)$	—	34	—
Hardness, Rockwell	—	R45–55	—
Water absorption, 24			
h (%)	0.01–0.52	0.4–0.55	—

inets. Many are cast or molded by RIM and RRIM techniques. (See Molding, Casting Processes.) Low-density flexible foams replace many latex rubber applications. Much is made from toluene diisocyanate (TDI) and PMDI for more-dense, less-flexible foams. Applications include crash pads, arm rests, door panels, carpet underlay, and nearly all furniture and bedding foam.

Polyurethanes behave as elastomers and are used where superior toughness and resistances to tear, abrasion, ozone, oxidation, and chemicals are needed. They are cast and molded by several techniques. With block copolymer blending of polyester or polyether, a number of high-performance applications are extended. The alternate blocks on the molecular chain provide a combination of desirable properties from each monomer. These polyurethane copolymer elastomers (PUR) are molded into solid tires, roller skate wheels, automotive door panels, bumper components, headlong housings, flexible fibers, electrical wire insulation, O-rings, hoses, and tubing. (See Figure 2-12.)

Polyurethanes find wide industrial applications as coatings. These coatings are tough, flexible, chemical resistant, and fast curing and have good adhesion. Various grades and formulations are used as interior and exterior finishes on wood, metal, and leather.

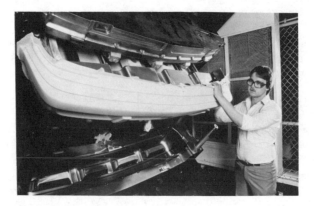

Figure 2-12. Reinforced polyurethane bumper fascias are produced by reaction molding (RIM) method. (Mobay Chemical Corporation)

Silicone (SI)

This family of polymers is derived from silica (sand) and methyl chloride or chlorobenzene. The word **silicone** should be applied only to polymers containing silicon–oxygen–silicon bonding. It is often used to denote any polymer containing silicon atoms. Silicone polymers are more correctly polyorganosiloxanes. This family does not have a carbon backbone characteristic of organic materials; consequently, they are sometimes called inorganic or semiorganic polymers. Various other elements with covalent bonding capacity, such as boron, aluminum, titanium, tin, lead, nitrogen, phosphorus, arsenic, sulfur, and selenium, have been studied for polymer production.

Various silicone structures—linear, cyclic, and branched—are used to produce fluids, resins, and elastomers. Silicones may be characterized as having a wide range of thermal capability, chemical resistance, and molecular structure (see Table 2-28).

Silicone fluids are used as antifoams, lubricants, paint additives, and release agents.

Silicone resins are formulated to develop into a three-dimensional network when fully cured. Most are used as low molecular weight solutions and cast with catalysts to form a condensation reaction. Applications include potting and encapsulation of electronic components, masonry treatments, paint additives, and adhesives. Lam-

Table 2-28. Selected Properties of Silicone

Properties	Mineral and/or Glass Filled	Silicone/Epoxy
Molding qualities	Good	Good
Relative density	1.80–2.05	1.84
Melting temperature (°C) T_m	Thermoset	Thermoset
T_g	—	—
Processing range (°F)	330–370	350
Molding pressure	0.3–6	0.4–1.0
Shrinkage	0.0–0.005	0.005
Tensile strength	4,000–6,500	6,000–8,000
Compressive strength	10,000–16,000	28,000
Flexural strength	8,000–14,000	17,000
Izod impact (ft-lb/in.) (1/8 in. specimen)	0.25–8.0	0.3
Linear expansion $(10^{-6}$ in./in./°C)	20–50	30
Hardness, Rockwell	M80–95	—
Flammability	V-0	V-0
Water absorption, 24 h (%)	0.15	—

inates reinforced with glass cloth find uses as ducts and electronic panel boards.

Silicone elastomers are made into two general types. Heat-curing elastomers are cured at elevated temperatures and made into tubing, O-rings, foams, fabric coatings, gaskets, hoses, and electrical connectors. Room-temperature vulcanizing (RTV) grades are formulated to cure by either a condensation or an addition reaction. Familiar condensation (one-part) RTV elastomers are used for caulking, adhesives, and making in-place gaskets. Contact with moist air promotes the curing reaction. In two-part systems the polymerization begins when the components are mixed. They are then used for potting and encapsulating electronic devices and for mold making. Faithful reproductions of intricate details may be copied from original patterns. The furniture industry uses silicone molds to reproduce wooden carvings and moldings.

Chapter 3.
What Additives Will Be Needed?

Introduction

In this chapter you will become familiar with the additives that enhance the characteristics of polymers and resin monomers. There are many ingredients that may be added to the polymers; only a few polymers are used in their pure form. You have seen how the properties of a single polymer may be radically altered by copolymerization and alloying. Resins and polymer compounds are also altered by chemical additives. Some of the reasons for including additives, reinforcements, and fillers are (1) to improve processability, (2) to reduce material costs, (3) to permit higher curing temperatures by reducing or diluting reactive materials, (4) to reduce shrinkage, (5) to improve surface finish, (6) to change the thermal properties such as expansion coefficient, flammability, and conductivity, (7) to improve electrical properties including conductivity or resistance, (8) to prevent degradation during fabrication and in service, (9) to provide desirable color or tint, (10) to improve mechanical properties such as modulus, strength, hardness, abrasion resistance, and toughness, and (11) to lower the coefficient of friction.

Some polymers could not be processed or fabricated into useful products if additives were not included. Not all additives may be used for every application. Some must be carefully selected and the amounts limited. The Food and Drug Administration (FDA) is concerned about substances capable of migrating into food or degassing with possible toxic effects. The National Fire Protection Association (NFPA) assists in disseminating standards intended to minimize the effect of fire. The Society of Plastics Engineers (SPE) and the Society of Plastics Industry (SPI) are also interested in issues of toxicology, combustibility, waste, and energy. (See Sources of Safety Information, Standards, Specifications, and Sources of Help.)

The processing or fabrication techniques along with quality control have a marked effect on the properties of plastics. Additives such as reinforcements may not have the desired effect if the processing technique is faulty. (See Molding Processes.) Voids, poor adhesion, excessive compounds, improper orientation, or other errors in processing and design may result in poor results. These problems are especially acute in composite structures and materials.

Additives

The following additives in resins and polymers are briefly outlined. A number of sources and suppliers list these additives and intended applications. (See Bibliography.) As a rule, additives do not comprise more than 10% (by weight) of the finished product.

There are many factors that must be considered before choosing an additive agent. (See Sources of Safety Information.)

1. Has the additive caused toxic or hazardous results from handling by or exposure to workers?
2. Has the additive been cleared by the FDA for use in food-contact applications?
3. Will there be a migration or deterioration of the additive over a period of time?
4. Will the additive be able to withstand the processing conditions involved?
5. Does the additive interfere with critical polymer properties such as heat sealability, modulus, chemical, electrical, or others?
6. Does the additive cause opacity or discoloration in application?
7. Do the benefits justify the cost?
8. Have the recommendations and specifications of competitive suppliers been compared for present application needs?
9. Will this additive have an impact on design and tooling parameters?

10. Is special handling, employee training, or auxiliary equipment needed or required?
11. Is the additive compatible with the host polymer and other additives?

Antioxidants

Antioxidants inhibit or retard oxidation during fabrication, storage, and service. Polypropylene and polyethylene are particularly susceptible to breakdown through oxidation. Deterioration may be controlled by absorbing or blocking ultraviolet light, deactivating metal ions, or decomposing hydroperoxides to nonradical products. **Antiozonants** are additives that help prevent breakdown by ozone gas in the atmosphere. Polyolefins, polystyrene, polyvinyl chloride, ABS, polyesters, and polyurethanes are susceptible to ultraviolet damage. Thermal, solar, and electrical radiation may be sufficient to break polymer bonds. This damage may result in crazing, chalking, color change, or degradation of physical, electrical, or chemical properties. **Ultraviolet stabilizers** are used in polyesters, polystyrene, acrylics, and polyethylenes. Carbon black may serve as a blocker of UV light in some polymers.

Antistatic

Those polymers that are not filled with conductive materials have a difficult time discharging ion (static) charges from their surfaces. Much of this charge is built up during processing, handling, or packaging. This charge results in the attraction of small particles, electrical discharges, and processing or handling cling. The major cause is friction. During processing, fabrication, and packaging operations, proper grounding and installation of an ionizing apparatus will help to neutralize these charges. An antistatic additive may be compounded into the polymer to attract moisture from the air, which in turn dissipates static charges from the surface because of the increased surface conductivity. A temporary external antistatic agent may be used, but it is easily removed in cleaning, handling, and service.

Colorants

Colorants are chosen for strength, electrical properties, specific gravity, clarity, and resistance to migration and degradation. There are four types of colorant used in polymers: (1) dyes, (2) organic pigments, (3) inorganic pigments, and (4) special-effect pigments. All four types may be compounded into a natural material and used as concentrates to provide better color control, facilitate handling, provide cost-effectiveness, and ensure safety. In addition to handling hazards, powders may be explosive. Most concentrates are made into a granular or pellet form and mixed with the base resin (plastics) prior to or during processing.

Dyes are usually soluble, strong, bright, transparent, and form chemical linkages with polymer molecules. Most dyes have excellent clarity but poor thermal and light stability. Some are dissolved directly by the resin monomer solution or added with selected solvents to the compounding of plastics and resins.

Organic pigments usually surpass inorganic pigments for brilliance and variety of hue. Like all pigments, they must be dispersed in the resin or plastics compound because they are not soluble. Particle size and dispersion affect color shade and strength. They are more costly, transparent, and susceptible to migration and thermal degradation than inorganics.

Inorganic pigments have larger particle size, better holding ability, and resistance to heat, light, and migration than organic pigments. In addition, they are generally lower in cost.

Special-effect pigments may be either organic or inorganic compounds and include fluorescents, phosphorescents, pearlescents, and metallics.

Coupling Agents

Composites, laminates, and other polymer compounds have one or more interfaces. The physical and mechanical properties are significantly affected by the structure and strength of the interface bond. This interfacial bond between the matrix, reinforcements, fillers, or laminates may be improved by surface treatment. Many resins and polymers do not want to adhere to glass fibers or other substrates. A number of coupling agents have been developed for the treatment of substrates. Silane and titanate coupling agents are used to promote adhesion between the matrix, fibers, particles, or laminar interfaces. Coupling agents are sometimes called promotors (not to be confused with promoters). Good adhesion between the matrix and reinforcements results in improved moisture, chemical, and heat resistances. Mechanical properties are often dramatically improved as the matrix transfers stress from one fiber to the next if there is a strong interfacial bond.

In some formulations the coupling agent may be added directly to the resin or polymer when compounding fillers, reinforcements, and other additives. Titanate agents improve strength and lower melt viscosity. Lower viscosity may result in reducing energy requirements during processing, while increasing production rates.

Silane coupling agents are commonly used as surface treatments for fibers, particles, or laminar substrates. It may be dry blended or applied in solvents or aqueous solutions. Adhesion is improved between sealants, adhesives, and plastisol coatings on substrates.

Flame Retardants

Polymers will support combustion, burn, produce smoke and toxic gases, and some will melt. They are organic carbonaceous materials. Flame retardants are designed to (1) insulate the material by forming a foam or char barrier, (2) create a cooling reaction by releasing water when heated, (3) inhibit burning by reducing the kind or amount of combustible material, and (4) interfere chemically with the combustion by releasing halogen or phosphorus compounds when heated.

Flame retardants may be internal or external additives. Smoke suppressants are also used to reduce the amount of particulates released when materials burn.

Foaming Agents

The term covers a wide variety of compounds and processing techniques. The terms **frothed, cellular, blown,** or **bubble** plastics are sometimes used. (See Expansion Processes.) These terms refer to the process of making polymers with a cellular structure. Cellular polymers may have the following advantages over solid polymers: cost savings, weight reduction, strength-to-weight ratio, thermal insulation, impact energy absorption, and a variety of densities with degrees of flexibility.

There are six methods of making polymers cellular:

1. Thermal breakdown of a chemical blowing agent. Various agents are added to the resin or polymer melt. When the chemical agent decomposes at a specified temperature, gases are released causing the cellular structure. Azodicarbon-

amide is widely used with HDPE, PP, ABS, PS, PVC, EVA, PMMA, and PPQ.

2. Dissolving or forcing a gas into the resin or melt that expands in the mold. Nitrogen is commonly used. This is sometimes called physical foaming.

3. Mixing a liquid component that vaporizes when heated. Volatile liquids (aliphatic hydrocarbons and chlorinated hydrocarbons) may be mixed with the resin or melt. The blowing agent is entrapped in the polymer and, upon heating, the expanding gases cause the softened polymer to fill the mold and stick together. Methylene chloride is used to expand flexible urethane polymers.

4. Whipping air into the resin and rapidly curing or cooling the polymer. The process is something like whipping hot wax.

5. Adding components that liberate gas within the resin by chemical reaction. Some resins may be mixed with catalysts, blowing agents, and other additives and forced into a mold. The chemical reaction causes polymerization and the cellular structure results from liberated gas.

6. Syntactic cellular polymers are produced by placing spheres or microballoons (25–150 μm) of glass, phenolic, urea–formaldehyde, or silica into a matrix. Polyesters, epoxies, and urea–formaldehyde are common matrices. Other novel ideas include mixing other cellular foams, pieces, or expandable beads into the resin mixture. (See Sintering.)

Heat Stabilizers

Stabilizers are added to polymers to prevent degradation caused by heat, light energy, oxidation, or mechanical shear. Stabilizers may include antioxidants, antiozonants, and ultraviolet absorbers. PVC, HDPE, PP, ABS, and PS may be stabilized with organophosphites. Metal salts of calcium/zinc are used in PVC because of their low toxicity. Flexible PVC food packages, tubing, or containers are typical applications. Organotin compounds are commonly used in rigid PVC products.

Impact Modifiers

This is a term given to polymers whose properties are modified by the addition of other monomers (usually elastomers). Impact modifiers of styrene–butadiene or polybutadiene are commonly grafted to acrylics. PVC is toughened by modification with ABS, CPE, EVA, or other elastomers. (See Ethylene–Ethyl Acrylate, Styrene–Butadiene.)

Lubricants

There are a number of reasons to include lubricants in polymers. They help lower friction between the polymer melt and the processing equipment and against the polymer itself (lower melt viscosity). Lubricants may aid in emulsifying other ingredients. Some lubricants prevent the plastics from sticking to the mold. Lubricants may include fatty acids, esters, fatty alcohols, metallic stearates, waxes, polyethylenes, silicones, or fluoroplastics. Calcium stearates are widely used in PVC, PP, ABS, polyesters, and PF.

All additives must be selected carefully for toxicological effects and desired service use.

Curing Agents

Curing agents are a group of chemicals that produce a cross-linked, thermosetting polymer from an initially linear or branched polymer or resin. Hexamethylenetetramine is used as a curing agent in B-stage novolac resins. When it decomposes in the presence of heat and moisture (a product of the condensation process), a cross-linked C-stage plastics is formed.

Because resins may be partially polymerized systems and other forms of energy may cause premature polymerization, **inhibitors** (stabilizers) may be used to prolong storage and block polymerization.

Accelerators (promotors) are additives that react in a manner opposite to that of inhibitors. They are mixed in the resin and speed up the chemical reaction between the inhibitor and resin. Accelerators are used to speed the decomposition of the initiator and subsequent cure. Many unsaturated polyester resins are supplied with the accelerator already added. Prepromoted polyester resins (resins with accelerator added) are commonly cured at ambient temperature. A common accelerator is cobalt naphthanate.

Catalysts, sometimes called hardeners (more correctly **initia-**

tors), cause the ends of monomers or other groups to join (polymerize) and/or cross-link. Organic peroxides are used as initiators in unsaturated polyester. Initiators are little influenced by the inhibitors.

There are a number of parameters to consider in selecting the proper initiator: shelf life, curing temperature, amount, resins system, gel time, processing technique, and application. The initiator benzoyl peroxide may be used when heat is used to aid in the full cure of the unsaturated polyester resin. The most common ambient initiator for polyester resin is methyl ethyl ketone peroxide. It reacts with the accelerator and attains full cure within a short period of time.

Organic peroxides may be used to polymerize thermoplastic materials. Monomer solutions are initiated and caused to become high-molecular weight polymer chains. Polyvinyl chloride, DLPE, PS, and PMMA are familiar examples. Some peroxides may be used to cross-link thermoplastic polymers such as LDPE, EVA, and HDPE.

All accelerators and initiators must be handled with care because they can cause skin irritation and chemical burns. They must never be mixed directly together because a violent reaction may occur.

Plasticizers

Plasticizers are additives used to increase flexibility, reduce melt temperature, and lower the melt viscosity, elastic modulus, and glass transition temperature of a polymer. Elasticizers act as an internal lubricant and weaken the intermolecular attraction of van der Waals bonds. Although they act like solvents, plasticizers are not designed to evaporate from the polymer during normal service life. Plasticizer leaching or loss is an important consideration. It is undesirable when in contact with food, pharmaceuticals, or other products for human consumption. They should be odorless, tasteless, nontoxic, and nonflammable. Leaching and degassing may cause PVC floor or automobile upholstery to become brittle and crack. Plasticized PVC hose becomes stiff or less flexible with plasticizer loss. Di-2-ethylhexyl phthalate (DOP) is a widely used plasticizer for PVC.

No one plasticizer can fulfill all requirements. There are more than 500 different plasticizers and blends used in polymers today. The plasticizer and polymer must have similar solubility parameters.

Preservatives

Although most polymers are not attacked by microorganisms, the additives such as plasticizers, lubricants, or other components may aid in microbiologic deterioration. Preservatives may be needed to prevent attack by microorganisms, insects, or rodents. Elastomers and heavily plasticized PVC are most susceptible. They are commonly subjected to wet, warm environmental conditions such as shower curtains, automobile tops, pool liners, awnings, cable coatings, boat coverings, or tubing. As moisture stays or condenses on the surface, growth media are collected. The EPA carefully regulates the use and handling of all antimicrobials. (See Bibliography, Personnel Hygiene.) Mildewicides, fungicides, and rodenticides may be used to provide the necessary protection in polymers.

Processing Aids

Processing aids are additives that prevent slip and adhesion and provide mold release and internal lubrication. The addition of another polymer may be considered a processing aid to improve processing behavior. Some of these aids may result in higher production rates or improved surface finish.

To prevent polymer compounds from adhering to processing equipment or molds, **mold release** agents may be applied to these surfaces. Waxes, silicones, fluoroplastics, and metallic stearates are common mold release agents. The antistatic agents such as fatty acid amines are also used as slip and antiblocking agents. These are used to prevent layers of plastic film or sheets from sticking to each other. **Slip** agents are added to the melt to give internal and external lubrication as they exude to the surface. **Antiblocking** agents such as waxes serve the same function and prevent two polymer surfaces from adhering. **Emulsifiers** lower the surface tension between compounds, thus acting as detergents and wetting agents. Emulsifiers are used in emulsion polymerization, food processing, and adhesives. Wetting agents used to lower viscosity are called **viscosity depressants**. They are used in plastisol compounds to assist in processing heavily filled materials or those that become too thick with age. Liquid ethoxylated fatty acids are found to be effective. **Solvents** may be considered a processing aid because they can be added to resins, dispersions, or polymers to dilute or assist in processing. (See Solvent Casting.) In addition, solvents are useful for cleaning resins from tools and equipment. (See Personnel Hygiene.)

Particulates

The term **particulates** (filler) is used to distinguish these materials from fibers (reinforcements) and to refer to a large, diverse group of materials that consist of minute particles. They may be classified according to source, function, composition, or morphology. Ambiguity of terms and overlap of function add to the confusion. Particulates are normally added to reduce costs, thus they are truly fillers. Fillers are generally not added as reinforcing agents, but some may significantly increase strength. Carbon black is a common reinforcing agent for elastomers. The geometry (morphology) of the particulate may include some fibers if the function was to lower cost. Natural organic fibers have been used as inexpensive fillers in urea and melamine plastics for years. Geometrics also vary greatly, with some being saucer, sphere, or needle shaped while others are irregularly shaped.

A number of other terms are used to describe these particulates in the matrix. The original meaning of the term **filler** was to describe any additive used to "fill" space in the matrix and lower cost. The word **extender** is sometimes used but may be inappropriate since some fillers are more expensive than the matrix. The term **dilutant** may be used in some literature to describe the particulate effect on curing or viscosity. The term **enhancers** is sometimes used to describe the addition of nonfibrous particulates to "enhance" properties.

According to ASTM, a filler is a relatively inert material added to a plastic to modify its strength, permanence, working properties, or other qualities or to lower costs.

Particulates are added to the polymer matrix for a number of reasons. Some have a synergistic effect (i.e., the combined effect is greater than the sum of individual effects). Particulates may reduce costs, lower the coefficient of linear expansion, reduce shrinkage, reduce molding cycles, increase thermal conductivity, lower resistivity, increase viscosity, slow the reaction of chemical additives, lower heat reaction (exotherm), lubricate, improve strength, reduce weight, improve flame retardation, suppress toxic smoke generation, increase hardness and stiffness, and reduce warpage. In general, elongation, ultimate strength, and toughness suffer. Some filaments abrade mold walls and feed paths.

Only two general classifications are used in this text: (1) particle morphology and (2) composition. It is appropriate to describe the function of some particulates in each classification.

Particles in polymers increase the number of variables that affect properties. Compounding these variables are the diversity of particle shapes, sizes, loading, packing, surface area, and density.

The importance of particle morphology should be apparent. During some processing techniques, particles in the matrix must be able to roll, slip, or tumble past each other and perhaps realign. Any nonspherical shape will result in greater viscosity or processing difficulty. Flakes or fibers will have aspect ratios (ratio of length to diameter of particle) that result in resistance to movement or realignment. Spheres have no aspect ratio. (See Figure 3-1.) They are easily moved in the matrix; consequently, they provide limited reinforcement properties. The mechanism of a filler in improving strength is to reduce the mobility of polymer chains. It may be expressed as a theoretical dimension. The degree to which a particle varies from sphericity may be expressed as a **sphericity coefficient**:

$$\text{sphericity} = \frac{\text{surface area of sphere with same volume as particle}}{\text{surface area of particle}}$$

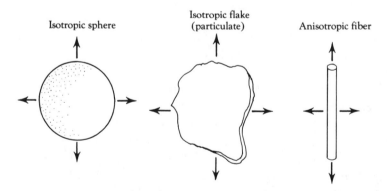

Isotropic sphere   Isotropic flake (particulate)   Anisotropic fiber

Figure 3-1. Spheres are isotropic but have no aspect ratio. Isotropic particulates have uniform mechanical properties in the plane of the flake. Fibers have lower aspect ratios but are anisotropic.

As the filler and matrix are compounded, an energy barrier may resist the wetting of the surface. All composite materials and structures depend on the interfacial bond between the filler and the matrix. The surfaces of fillers are commonly treated with a coupling agent. (See Coupling Agents.)

The size and amount of filler in the matrix greatly influence composite properties. The amount of filler is often termed **loading** and is expressed quantitatively. Predictive models based on the packing geometry of the particles are useful in obtaining optimum loading and the desired functional improvements in the composite. As a rule, large particles (>40 μm) tend to settle, and there is a lot of space between particles. If matrix economy is important, **particle size distribution** must be considered. If a minimum amount of matrix is to be used, a fairly wide range of particle sizes must be used.

If all particles were of the same size sphere, nearly half of the composite would be composed of matrix. It is reasonable to expect that maximum loading would result if smaller particles fit between the medium and the larger particles. Pigments, or other additives in the matrix, complicate the computer model of filler loading. Very high loading with relatively little rise in viscosity is permitted because of the ballbearing action of spheres. The introduction of different shapes and sizes may result in a synergistic effect. One layer of flakes may have an anisotropic property while the spheres on another layer would be isotropic.

Relative **density** may be an important consideration in some applications. Glass and plastic hollow spheres will result in lower composite density, which may be most beneficial where weight (mass) savings are critical. They are incorporated into SMC to make automotive and aerospace parts. Low-density filled composites may find applications where improved thermal and acoustical insulations are needed. Both solid and hollow spheres reduce shrinkage and in clusters (like grapes) provide some aspect ratios for improvement in tensile strength.

Flakes and fiber fillers will have aspect ratios that make them useful in reinforcing. Both tend to orient with matrix flow and processing. This may be advantageous in thermoformed, injection-molded, or extruded parts. Flakes are easily broken during processing and the sharp edges may also result in high stress concentrations. Moisture, weather, or chemicals may cause stress cracking on the surface, resulting in further deterioration. Because flakes are flat, they may be closely packed to provide a high percentage of reinforcing for a given cross-sectional area. They form a barrier, like fish scales, that may stop passage of vapors and reduce the possibility of deterioration of the composite by penetration.

Composites requiring electromagnetic interference shielding may use metal or metal-coated fillers or flakes.

There are two classifications of fillers by composition: (1) inorganic and (2) organic. **Inorganic** fillers are nonmetallic minerals such as kaolin, talc, calcium carbonate, and metallic fillers consist of such metals as aluminum, bronze, and zinc. **Organic** fillers include wood, natural fibers, and some polymers. Selected fillers and reinforcements with functions are shown in Table 3-1. This table does not indicate the degree of improvements of function. The prime function will also vary between thermosetting and thermoplastic matrices. This table is to be used only as a guide to the selection of fillers. Inorganic thixotropic fillers increase viscosity and act as emulsifiers to prevent separation of additives in the matrix. Thixotrophy is a state of a material that is gellike at rest but fluid when agitated. This is most desirable in paints, BMC, and SMC. Fumed silica is a common thixotrophic filler. Calcium carbonates and kaolin (ingredient of clay) are commonly used in BMC, SMC, and other composites because of low costs and relatively low abrasivity. Asbestos and wollastonite are fibrous minerals that fill and reinforce brake linings and other products. Conductive composites are obtained by the addition of metal fillers or metal-coated fillers.

Wood flour is widely used as an organic filler in phenolics and other thermosets. Cellulosic products and polymer fibers may degrade or decompose upon processing. Carbon and graphite are now among the most widely used organic fillers. They are used as reinforcements and conductive agents. Carbon black is used to improve the UV radiation resistance in polyolefins.

Fibers (Reinforcements)

Reinforcements are often confused with fillers because some materials (glass, asbestos) may act as a reinforcing agent or filler. As a rule, the reinforcements are fibrous. These fibers produce a dramatic improvement in the physical properties of composites. Composites reinforced with fibers are anisotropic, and properties are dependent on the direction of the stress.

A fiber is, by definition, something different than a particle. Its length is greater than its equivalent cylindrical diameter, and the fiber is a better reinforcer as a result. Many believe that "more is better" with respect to aspect ratio. In fact, this is not necessarily so. Fibers have large aspect ratios (length-to-thickness). As the ratio of fiber length to fiber diameter increases, the strength of the composite increases. Most of the stress in fibers is transmitted by the matrix to the fiber ends. Properties of commonly used fiber reinforcing agents are shown in Table 3-2.

There are a number of variables that influence the properties of fibrous reinforced composite materials and structures: (1) interface bond between matrix and fiber, (2) properties of the fiber, (3) size and shape of the fiber, (4) loading of fiber, (5) processing technique, and (6) alignment or distribution of the fiber.

Interface Bond

Fibers are as strong as their weakest point. Any defects such as voids, cracks, nicks, or other discontinuities will be influenced by the fiber diameter. The finer the fiber is, the smaller the defect. Most fibers are very strong when pulled, but will bend and buckle under low forces when pushed axially. It is the function of the matrix to prevent fibers from bending and buckling. If the matrix must keep these fibers from moving and transfer stress, there must be excellent adhesion between the matrix and the fiber. (See Coupling Agents.) Organosilanes are widely used as primers on glass fibers to promote adhesion. One reason that resinous phase polymers are popular is that they wet the fiber surface. They also require little processing energy to change from this liquid to a solid phase. Processing equipment may be less elaborate, and larger components may be manufactured.

If nearly every matrix is made stronger by the addition of fiber reinforcements, the fiber composition must be vital. There are a number of fibrous classes: (1) glass, (2) carbonaceous, (3) polymer, (4) inorganic, (5) metal, and (6) hybrids.

Glass Fibers

The term **reinforced plastics** is used primarily to describe glass fiber reinforced plastics (FRP). Glass fibers have been popular reinforcing agents for more than 40 years. Most of the early applications

Table 3-1. Selected Fillers and Reinforcements with Function

Function	Filler/Reinforcement																
	Glass fiber	Asbestos	Wollastonite	Carbon Fiber	Whiskers	Synthetic Fibers	Cellulose/ Wood	Mica	Talc	Graphite	Sand/ Quartz Powder	Silica	Clay	Glass Spheres	Calcium Carbonate	Metallic Oxides	Carbon Black
Hardness	+	+	+					+			+	+	+	+			
Tensile strength	+	+		+	+	+		+	○	+				+			
Compressive strength	+		+						+		+	+	+	+	+		
Modulus of elasticity	+	+	+	+	+	+		+	+		+	+		+	+	+	+
Impact strength	±	−	−	−	−	+	+	±	−		−	−	−	−	±	−	+
Thermal expansion	+	+	+	+	+			+	+		+	+	+			+	
Shrinkage	+	+	+	+				+	+	+	+	+	+	+	+	+	+

Thermal conductivity	+	+	+	+					+	+	+	+			+	+	+
Bulk	+	+	+				+	+	+	+	+	+			+	+	+
Electrical conductivity	+			+						+						+	+
Electrical resistance	+	+	+					+	+			+	+			−	
Thermal stability	+	+	+					+	+		+	+	+			+	+
Chemical resistance	+		+					+	o	+			+	+			
Abrasion behavior	+			+				+	+	+			+				
Moisture resistance	+	+	+					+	+		+	+	+	+	+	+	
Machine absorption	o				o	o	o		o	o	−	−		o	o		o
Price reduction	+	+	+	−	−	−	+	+	+	+	+	+	+	+	+	−	+
UV radiation	+	+	+				+	+	+						+	+	−

+, positive influence; o, no influence; −, negative influence.

Table 3-2. Properties of the Most Commonly Used Fiber
Reinforcing Agents (Metallic and Nonmetallic)

Fiber	Relative Density	Melting Point (°C)	Ultimate Tensile Strength (MPa)	Tensile Modulus of Elasticity (GPa)
Aluminum	2.70	660	620	73
Aluminum oxide	3.97	2082	689	323
Aluminum silica	3.90	1816	4130	100
Asbestos	2.50	1521	1380	172
Beryllium	1.84	1284	1310	303
Beryllium carbide	2.44	2093	1030	310
Beryllium oxide	3.03	2566	517	352
Boron–tungsten boride	2.30	2100	3450	441
Carbon	1.76	3700	2760	200
Glass, E-glass	2.54	1316	3450	72
S-glass	2.49	1650	4820	85
Graphite	1.50	3650	2760	345
Molybdenum	10.20	2610	1380	358
Polyamide	1.14	249	827	2.8
Polyester	1.40	249	689	4.1
Quartz (fused silica)	2.20	1927		70
Steel	7.87	1621	4130	200
Tantalum	16.60	2996	620	193
Titanium	4.72	1668	1930	114
Tungsten	19.30	3410	4270	400

were confined to using thermosetting resins. Today, both thermosetting and thermoplastics are used to produce glass-reinforced composite products. Glass-reinforced composites may have high strength-
to-weight ratios; dimensional stability; and resistance to heat, cold,
moisture, and corrosion. There are several compositions of continuous fiber glass. The major constituent is silica, with other additives
added to produce a variety of properties.

It has been estimated that more than 95% of the glass fiber reinforcements are made for **E-glass** (electrical glass). This glass exhibits good electrical characteristics. It has tensile strength of
200,000–300,000 psi with a relative density of 2.55. E-glass con-
stitutes the majority of glass textile production.

If stiffer, stronger fiber is needed, **S-glass** (high-strength glass)
grades with tensile strength of about 650,000 psi and a density of

Table 3-3. Selected Properties of Glass Fibers

Property	A, High Alkali	C, Chemical Resistant	D, Low Dielectric	E, Electrical	S, High Strength
Relative density	2.50	2.49	2.16	2.54	2.49
Tensile strength (MPa) at 20°C	3,033	3,033	2,420	3,450	4,820
Modulus of elasticity (MPa) at 20°C	—	69,000	—	72,500	85,500
Coefficient of thermal linear expansion (10^{-6} in./in./°F)	4.8	4.0	1.8	2.8	3.1

2.48 are used. This glass is about 20% stronger and stiffer than E-glass. It is usually about five times more costly. Most applications are for structural composites in the aerospace industry.

A soda-lime glass (high alkali) known as **A-glass** is commonly manufactured into containers requiring chemical resistance to acids.

Where chemical resistance is of primary concern, **C-glass** (chemical glass) is selected.

For applications requiring a low dielectric constant and low density, **D-glass** (dielectric glass) may be selected. Radomes are typical electronic applications.

A glass with lead oxide for improved radiation protection is **L-glass** (lead glass). It may be used in x-ray protective clothing or as a tracer fiber to verify fiber alignment in other composite mixtures.

Some selected properties of fibrous glass are shown in Table 3-3.

Carbonaceous Fibers

Most carbon fibers result from heating and controlling the temperature, tension, and atmosphere of the organic fiber in a pyrolysis chamber. Once all organic materials have been driven off, the result is a high-strength, high-modulus, low-density fiber. An early organic precursor (one that precedes the other) was Rayon. It was pyrolized and stretched into filaments to produce an oriented fiber. Polyacrylonitrile (PAN) is also used as a precursor to produce high-performance carbon and graphite fibers. Ordinary pitches are isotropic in character and must be oriented to be useful as reinforcing agents. (See Figure 3-2.)

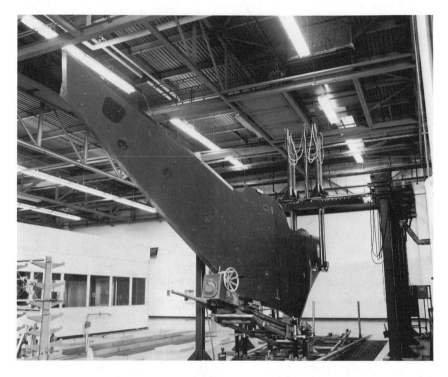

Figure 3-2. Carbon/epoxy composites are used for the stabilizer and wing structure (pictured) in the AV-8B Fighter. (McDonnell Aircraft Company)

Polymer Fibers

Almost any organic fiber may be used in a composite structure. For years, natural fibers (cotton, flax, silk) were used as the reinforcing agents with thermosetting polymers. These composites covered our airplanes, provided strength to tires, and were used as electrical insulation boards and gears. Many of these natural fibers have been replaced by synthetic materials.

Many polymer fibers are used as reinforcements with elastomers (rubber matrix).

Cellulosics, polyamides, polyvinyls, polyacrylonitrile, and polyesters have been used successfully in the preparation of strong composites. Many thermoplastic fibers improve impact strength, electrical properties, and chemical resistance in molded composites.

Special sizing systems enhance the cohesion (integrity) and wet-out characteristics of these fibers. Bundles of these fibers are cut and compounded into BMC, SMC, and TMC systems.

Aramid fiber is the generic name for aromatic polyamide fibers that have nearly twice the stiffness and about half the density of glass. The fiber was developed for use in the manufacture of radial tires. Two familiar aramid fibers produced by DuPont are Kevlar 29 and Kevlar 49. Kevlar 29 fibers are used for ballistic protection, ropes, coated fabrics, and as replacement for asbestos. Kevlar 49 is used in boat hulls, flywheels, V-belts, hoses, and composite armor.

Epoxy matrix is most common; however, polyesters, phenolic, and other resin systems are used. (See Figure 3-3.)

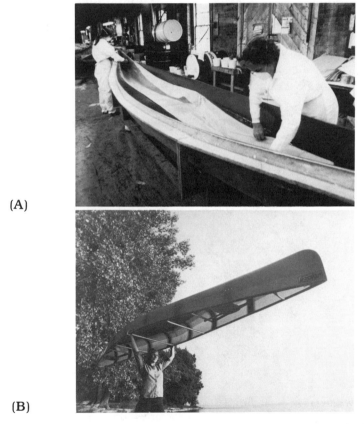

(A)

(B)

Figure 3-3. (A) Hand lay-up of Kevlar/epoxy canoe. (B) Kevlar/epoxy canoe is tough and light. (Hexcel)

Inorganic Fibers

Much of the development of inorganic fibers came about as a result of developing improved structural materials that would reduce weight and increase performance. Glass composites accomplished much of this but could not be considered for load structures because of their lack of stiffness.

A number of short crystalline fibers were developed with outstanding strength. They were called **whiskers**. Whisker crystals have been made from alumina, chromia, boron carbide, silicon carbide, and other materials. Only a few have found commercial success. (See Figure 3-4.) There are a number of problems associated with whisker composites. Production costs are very high and handling during processing results in additional costs. Potassium titanate whiskers are used in large quantities to strengthen composites in thermoplastic matrices (e.g., polyamide and acetates).

Most of the inorganic fibers used today are of continuous, mass-fabricated types. These include silicon carbide, boron carbide, titanium boride, or coating of these materials on other substrate filaments.

The most important inorganic continuous fiber is boron. It is

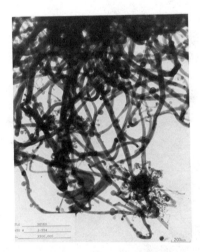

Figure 3-4. Whiskers of crystalline silicon surrounded by a sheath of silicon dioxide. They have an average diameter of about 0.05 μm. Classed as a "cobweb" whisker, they occur as individual fibers or random, disoriented bundles. (J. M. Huber Corporation)

stronger than carbon and may be incorporated into a polymer or aluminum matrix. Like whisker fibers, these materials have a relatively high cost. They may be produced by chemical vapor deposition on a tungsten substrate. Boron fibers may be coated with silicon carbide or boron carbide to provide improved performance in aluminum and titanium metal matrix composites. Boron–carbon filaments are less costly than boron–tungsten fibers. Most of the boron composites are used to make the tail assemblies of F-14 and F-15 aircraft. Epoxy matrix in preimpregnated boron tapes and cloth are commonly laminated into composite structures.

The tensile strength of boron–tungsten fibers may be improved by relieving the residual stresses at the inner surface of the boron. This is accomplished by etching away part of its outer surface.

Silicon carbide fibers are made and used in both polymer and aluminum matrices. Since boron will degrade when attacked by the chemical action of molten aluminum, boron fibers may be coated with diffusion barriers to overcome much of the unfavorable reaction with metal or other superalloy composites. Boron nitride may be used to coat boron, and tungsten carbide and tantalum carbide are used as diffusion barriers on silicon carbide.

Metal Fibers

Metal fibers are drawn from continuous strands of steel and other metals. Most are used in laminar applications in elastomers (rubber) or metal matrices. Steel, aluminum, and other metal fibers cannot compare with the strength and density exhibited by other fibers. A number of composite laminar structures use metal foils or plies. (See Hybrid, Laminar.)

Hybrid

Hybrid fibers are actually an available form of combining two or more reinforcing materials. When these fibers are placed in a matrix, the composite (not the fibers) is called a **hybrid**. The idea seems obvious but so does the idea of alloying and reinforcing any material. Hybridization allows the designer to tailor the composite for maximum performance and minimum cost. Generally, one component lowers cost while the other is added to enhance the deficiency of the other.

The diversity of properties and the possible material combinations are too numerous to detail. There are five major types of hybrid composite:

1. **Random** refers to the randomly mixed fibers in the matrix with no preference to major loading.
2. **Interply** is descriptive of placing two or more layers of fiber, cloth, or tape in a laminar composite. There is only one type of class or reinforcement, but there may be several discrete layers of this reinforcement.
3. **Intraply** is the combining of two or more reinforcements in a layer. They may be fiber, cloth, tape, or combinations of each.
4. **Selective placement** is what the name implies. Reinforcements are placed where additional strength is needed, over the base reinforcing laminate layer. Ribs, corners, edges, or other areas needing a composite laminar stiffener are most common.
5. A composite composed of metal foils or metal composite plies stacked in a specified orientation and sequence is referred to as a **superhybrid**. (See Laminar.)

The need for lightweight structural materials at lower cost has prompted many of these hybrid formulations. Glass fibers are heavy, have good impact strength, and are inexpensive. By combining them with carbon into a composite laminate, the carbon reduces weight and stiffness. There are some hybrid combinations that present an apparent synergistic effect on some properties. Graphite fibers may enhance the thermal and electrical properties of the hybrid, while carbon fibers may help in shielding applications, reduce static electricity, and assist in electrostatic coating operations.

Hybrid reinforcements of glass and polyester fibers are providing characteristics that combine the benefits of both fibers.

Properties of the Fiber

Composition is the primary determinant of properties for each fiber. The ways in which it is produced, handled, processed, surface enhanced, and hybridized, as well as the size, shape, and number of defects, all play an important role. Table 3-4 illustrates selected properties of common fibers. These figures are to be used only as a

Table 3-4. Selected Filament Properties

Filament	Density (kg/m³)	Tensile Strength (GPa)	Specific Strength (MPa-m³/kg)	Modulus (GPa)	Specific Modulus (MPa-m³/kg)	Elongation (%)
E-glass nonfilament	2550	3.45	1.35	72.4	28.4	4.8
S-glass nonfilament	2500	4.59	1.84	96.9	38.9	5.3
Graphite Thornel (Union Carbide)	1770	2.24	1.26	234	132	1.0
Aramid Kevlar 49 (DuPont)	1400	2.76	1.92	131	91	2.0

comparison; some values shown are from different testing procedures.

Size and Shape of Fiber

Fibers and filaments are produced in a number of different profile shapes, sizes, and forms. The term **fiber** is used to describe any single filament that is long and slender.

Some fibers are made in a variety of cross-sectional shapes (e.g., round, triangular, dogbone, ribbon, and lobular). Some may aid in providing superior handling, loading, processing, packing, orientation, or adhesion in the matrix.

With the exception of single-crystal whiskers, the full length of filament reinforcement may be used in filament-wound and pultruded composites. Some of the strongest composites are made from continuous filaments impregnated with resin before curing. (See Molding Processes.)

Some of the fibers are so small that they are handled in bundles. Others are woven into various cloth or tape orientations to ensure a strong anisotrophy of mechanical and physical properties. Carbon fibers are available in continuous lengths up to 40,000 ft in bundles (called tows) of 1000–160,000 filaments. They are about the same diameter as glass filaments (0.0003 in.). The filament diameter designations for glass fiber are shown in Table 3-5.

Short fibers, flakes, or particulates are more likely to be randomly distributed than long fibers. To obtain a variety of processing options and be able to tailor the performance of the finished composite, a number of reinforcing forms are available.

Table 3-5. Glass Fiber Designations

Filament Designation	Filament Diameter (μm)
B	3.80
C	4.50
D	5.00
DE	6.00
E	7.00
G	9.10
H	11.12
K	13.14
M	15.17
T	23.24

Fibrous reinforcements are supplied in several basic forms, including (1) rovings, (2) chopped strands, (3) mats, (4) yarns, (5) woven fabrics, (6) woven rovings, (7) nonwoven fabrics, and (8) tapes and other special configurations.

Rovings refer to a bundle of continuous filaments either as untwisted strands or as twisted yarns. In this form the reinforcing fibers may be wound on mandrel shapes, preformed, pultruded, or cut and applied where needed. Carbon, glass, thermoplastic and hybrid aramid/glass, and carbon/glass are used in this form. (See Figure 3-5.)

Chopped strands are generally rovings that have been cut in lengths from 3 to 50 mm. Much is used in premix molding, such as BMC or SMC wet slurry preforming, or in hand lay-up operations. **Milled** fibers are less than 1.5 mm in length and are produced by hammer milling the glass strands. Glass is used because of lower cost but it is more dense than many other fibers. Milled glass is useful as anticrazing reinforcing fillers. (See Figure 3-6.)

Reinforcing **mats** are continuous or chopped strands of randomly oriented or swirled filaments with a binder. They are produced in various widths, weights, and lengths and are held together by a resinous binder or by mechanical stitching called **needling**. Glass mats are used in matched die molding, hand lay-up continuous panel molding, and centrifugal casting operations.

Figure 3-5. Continuous strand roving. (PPG Industries)

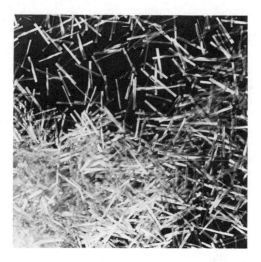

Figure 3-6.   Chopped fiberglass strand. (PPG Industries)

Yarns are a group of twisted fibers that can be woven into materials or filament wound on mandrels. Yarn nomenclature for glass is used to help specify the desired design properties of the completed composite. For example, a yarn designated as ECG 150 1/2 would be

E = electrical glass
C = continuous filament
  (T = texturized or fluffed)
G = filament diameter of 9 μm
  (from Table 3-5)
150 = 15,000 yd/lb
1/2 = single strands were twisted
  and two of the twisted strands
  were plied together (S or Z may be
  used to designate the type of
  twist)

There were two basic strands in the yarn. Thus,

$$\frac{15,000}{2} = 7500 \text{ yd/lb of yarn}$$

Tex is a unit of fiber fineness indicated by the weight in grams per kilometer. (See Table 3-6.) The lower the number, the finer the fil-

Table 3-6. Glass Fiber Yarns

Glass System				yd/lb	TEX
ECD	1	800	1/0	180,000	2.75
ECD	18	000	1/2	90,000	5.5
EDC		900	1/2	45,000	11
EDC		450	1/2	22,500	22
EDC		450	2/2	11,250	44
EDC		225	1/2	11,250	44
EDC		225	2/2	5,625	88
ECG		150	1/0	15,000	33
ECG		150	1/2	7,500	66
ECG		150	2/2	3,750	132
ECG		75	2/2	1,875	264
ECDE		37	1/0	3,700	134
ECK		18	1/0	1,800	275
SCD		450	1/2	22,500	22
SCD		150	1/2	7,500	66
ETDE		150	1/0	14,200	35
ETDE		75	1/0	7,100	70

ament or yarn. At one time, the fineness was expressed by a unit called denier, which is equal to the weight in grams of 9000 meters.

The term **woven fabrics** refers to a number of cloth weave patterns. The **warp** yarns run in the direction of the fabric (lengthwise) and the filler or **pick** (weft) yarns generally run crosswise or at right angles to the warp. Some of the varieties of weave pattern include plain, basket, twill, crowfoot satin, 8-shaft satin, and unidirectional. Graphite, boron, or hybrid fibers may be used in combinations with glass or woven with selected warp and pick patterns for the designed performance property.

Fabrics are commonly made into composite components by hand lay-up, vacuum bag, autoclave, pressure laminating, or other techniques. Some are made into prepregs. Prepregs are materials that have been impregnated with partially cured resin (B-stage). The fabricator then places the prepreg in the mold for final cure from the application of heat and pressure.

One manufacturer provides a knitted "matrix" reinforcement fabric. Although all the glass fiber is in the weft direction, a lightweight, high-tenacity, chemically activated yarn inserted in the warp direction allows the roving to be shifted during molding or pulled through pultrusion dies without breaking. There is less "kink stress" than in conventional woven materials. The concept is shown in Figure 3-7.

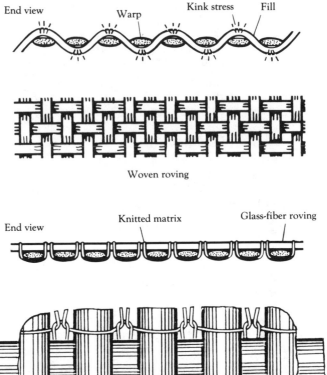

Figure 3-7. Schematic showing the source of "kink stress" in conventional woven fabric compared with knitted style. (Composite Reinforcements Business)

Figure 3-8. Woven roving fabric. (The CertainTeed Corporation)

Woven roving is a heavy fabric made with roving. These heavy coarse cloths are used mainly in hand lay-up operations. (See Figure 3-8.)

Nonwoven fabrics are made by using an adhesive to hold the warp and pick yarns together. There is no weaving or interlacing of yarns. The yarns are mostly unidirectional and generally stiff. They do not readily conform to complex shapes and are used in producing composite panels, large containers, braces or other structures. Continuous laminating, hand lay-up, and filament winding are common production techniques.

Tapes, contoured fabrics, fluted-core fabrics, and three-dimensional forms are also produced from fibers. Some may be woven into a tube shape with special reinforcing warp ribs. The cloth may be woven into a symmetrical shape (cap) for the production of domes or cone composites. Tapes are most common. They are usually less than 150 mm wide and may be made in several weave patterns.

Loading of Fiber

Generally, mechanical strength of the composite depends on the amount of glass or other reinforcing agent it contains. This assumes that the fiber has greater strength than that of the matrix. A part containing 60% fiber reinforcement and 10% resin is almost six times stronger than a part containing the opposite amounts of these two materials.

Closely related to the amount of filaments in the matrix is the packing or alignment of the fibers. You may place more filaments in a given volume if the arrangement is carefully planned. Glass content of up to 80% by weight can be achieved by unidirectional orientation. Typical fabric weave patterns of glass, carbon, aramids, and most hybrids are 70% by weight fiber and 30% resin. Most reinforced thermoplastic composites contain less than 40% by weight reinforcements.

These materials are commonly molded into isotropic composite components. Some processing techniques allow the reinforcements to be aligned or oriented more carefully, thus resulting in a more anisotropic composite dependent on the stress direction.

Carbon, boron, and glass loading in continuous filament winding processes allows for close packing. The result is a low resin content (20% by weight). Helical patterns result in higher resin content with less than optimum strength. Overlapping results in greater resin voids or interstices.

Processing Technique

Processing technology can assist us in designing composite products with desirable performance properties. There are a number of problems that may be attributed to present processing technology. We must realize that the molecular structure of the matrix will not be constant during the lifetime of the product. Heat, light, chemicals, and other agents may continue to degrade the matrix. However, not all polymer matrices are equally susceptible to these effects and selection is important. The composite is dependent on the ability of the matrix to transfer stresses. Adhesion and modulus are the two most important properties that control the ability of the matrix to transfer stresses.

During processing and handling of reinforcing fibers, it is imperative that fibers are given the correct surface treatment directed toward the intended processing method, resin compatibility, and end-use performance. (See Coupling Agents.) These chemical surface treatments help ensure a strong adhesion between the fibers and matrix.

Present technology does not allow us to produce fibers that are very close to theoretical strength. (See Whiskers.) Dislocations, grain boundaries, cracks, and voids are fiber defects that substantially affect strength. These defects are present in manufactured fibers and may be masticated by improper handling and processing techniques.

In processing, the molding methods must be carefully controlled. Excessive molding or curing temperatures may degrade the polymer matrix, fiber, or both. Poor molding designs and technique may result in stresses being molded into the composite. Warpage and other dimensional instabilities will be the result.

Remember, the strength of fibrous composites relies on the transfer of stresses from the matrix to the fibers. During processing some will be damaged and some will be stressed more than during molding. Fiber length is an important parameter in determining the stress to be transferred. This may be expressed as

$$L_f = DST_f$$

where

L_f = critical fiber length
D = diameter of fiber
S = strength of matrix bond to fiber (approximately equal to shear strength of matrix)
T_f = tensile strength of fiber

The strength of an aligned short-fiber composite decreases as the angle between the fiber axis and loading direction increases.

The ultimate tensile strength of many fibrous-reinforced composites may be estimated as

$$T_u = V_f \left(1 - \frac{L_f}{2L} \right) T_f + V_m T_m$$

where

T_u = ultimate tensile strength of composite
V_f = volume fraction of glass fibers
L_f = critical fiber length
L = length of fiber
T_f = tensile strength of fiber
V_m = volume fraction of matrix
T_m = tensile strength of matrix

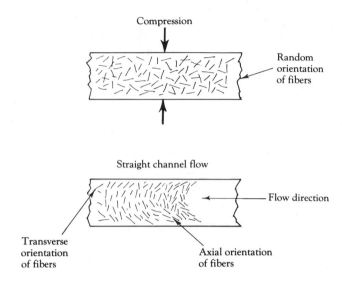

Figure 3-9. Fiber orientation in compression and straight channel flow.

It should not be surprising that the strength of molded composites falls short of the values expected for short or long reinforcements. The L_f for many fiber matrix formulations is about 2000 μm. Processing and manufacturing variables have a significant effect on the extent of fiber breakage, dispersion, length, orientation, and wet out adhesion.

In compression and injection processes, placement of gates, section thickness, vents, and cores can dramatically alter fiber orientation. Gating is perhaps the most important of the tooling design. In Figure 3-9 the compression of fibrous materials caused the fibers to align perpendicular to the mold closing resulting in a random orientation. In straight channel flow, fiber alignment parallel to the flow direction is likely to form. The concept of diverging and converging flows is shown in Figure 3-10.

If compounding or plasticating (hot mixing) processes result in excessive numbers of fibers being broken or damaged, mechanical properties may be lowered. Both thermoplastic and thermosetting polymers must fully wet the reinforcements for ultimate strength.

Much of the composite industry for aerospace structures is labor intensive. Hand lay-up methods and other techniques depend extensively on the ability and skill of human workers. Filament pack-

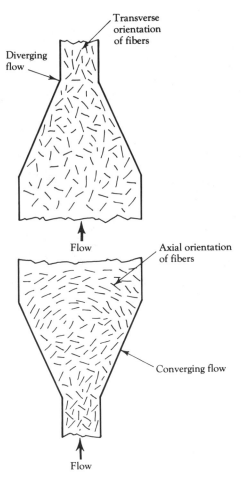

Figure 3-10. Diverging and converging flow illustrating fiber orientation.

ing, alignment, avoiding resin-rich pockets, and minimizing voids depend on human endeavor. Automated and high-volume production techniques also rely on operators for control monitoring and planned processing changes. (See Figure 3-11.)

It is a sophisticated combination of materials whose characteristics are determined as much by fabrication procedures as by properties of component materials. Composite materials can be made quite anisotropic through directional control of fibers in a weak matrix.

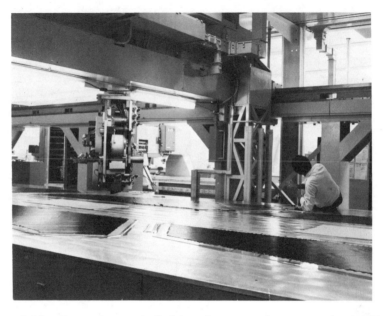

Figure 3-11. Computer-controlled tape-laying machine is used to build up composite F-16 fighter stabilizer components. Hand lay-up is slow and may result in parts with porosity, voids, and delamination defects. (General Dynamics—Fort Worth Division)

Alignment of the Fiber

Fibers cannot be used on their own for structural applications: They have to be embedded in a matrix. In many objects, stresses are not uniformly distributed, but careful alignment of the fibers can help to support the stress. In fishing rods, vaulting poles, and golf club shafts, the fibers are arranged along the long axis to resist the applied bending stresses. If a tension was applied in the direction of alignment, as shown in Figure 3-12, not all fibers would break at the same time. The strengths and moduli of elasticity of the fibers vary. In this case, the modulus of these composites can be estimated by the rule of mixtures: The total stress in the composite is equal to the sum of the stresses in its constituents and these individual stresses are proportional to their moduli of elasticity and to their relative volumes. The average modulus of elasticity is proportional to the moduli of elasticity of the constituents multiplied by the cross-sectional areas of those constituents and divided by the total cross section.

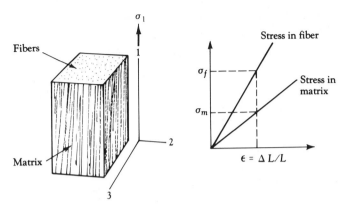

Figure 3-12. *Left:* Unidirectional reinforcement of composite with assumption of equal strain in fiber and matrix when stressed. *Right:* Stresses in fiber and matrix at equal strains are proportional to slope of stress–strain curves.

The density of concrete is the sum of the individual constituents (weights) of sand, gravel, and cement of which it is composed.

$$E_c = E_m V_m + E_f V_f$$

where

$$c = \text{concrete}$$
$$m = \text{matrix}$$
$$f = \text{fiber}$$
$$V_m = \text{volume fractions of the matrix}$$
$$V_f = \text{volume fractions of the fiber}$$

For a polymer matrix with no reinforcement, the equation reduces to

$$E_c = E_m (1) + E_f (0)$$
$$E_c = E_m$$

For reinforcements with no matrix, the equation reduces to

$$E_c = E_m (0) + E_f (1)$$
$$E_c = E_f$$

If the alignment is at right angles or has another distribution of the fiber, the rule of mixtures is no longer applicable.

Data taken from Table 3-7 may be used to calculate the theoretical modulus of this composite for any volume fraction of fiber. Using

Table 3-7. Properties of Carbon Fiber and
Epoxy Polymer

Resin	Epoxy
Density	1.3 g/cm³
Tensile strength	88 MPa
Tensile modulus	3.5×10^3 MPa
Tensile elongation to break	5.2%
Carbon fiber density	2.01
Carbon fiber tensile strength	2100 MPa
Carbon fiber modulus	390×10^3 MPa
Carbon volume fraction	0.65

the rule of mixtures formula, we can calculate the modulus of the composite in the direction of the fiber:

$$E_{c_1} = E_m V_m + E_f V_f$$
$$= (3.5 \times 10^3)(0.35) + (390 \times 10^3)(0.65)$$
$$= 254.725 \times 10^3 \text{ MPa}$$

Parallel alignment is found in pultrusion and some filament-wound applications.

To help illustrate that the stiffness of this composite in the other two directions (shown in Figure 3-12) is much less, the following calculation is performed:

$$E_{c_2} = E_{c_3} = \frac{E_f E_m}{E_f V_m + E_m V_f}$$

In Figure 3-13 the parallel (anisotropic) alignment of continuous strands gives the highest strength, bidirectional (cloth) alignment gives a middle strength range, and random (mat or isotropic) alignment gives the lowest strength range.

In bidirectional alignment, where half of the strands are laid at right angles to the other half (0° to 90°), the mechanical strength at either angle is less than that at parallel alignment. As this angle varies from 0° to 90°, the strength varies proportionally. When half the strands are laced at right angles to each other, loadings range from about 50 to 75%. Parallel loading can reach 85%, while loading in random alignment may exceed 50%. Remember, the matrix must keep the fibers from buckling and must transfer the stress to the surrounding fibers. Since the composite is being stressed in one direction and there are only a limited number of fibers in the one, two, and three direction, we must assume all will be lower strength.

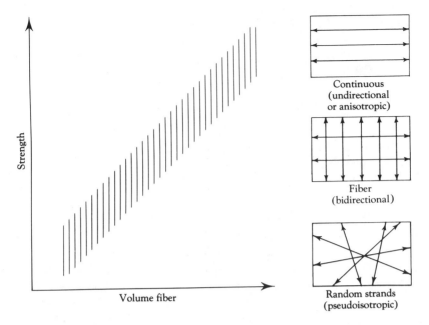

Figure 3-13. Strength relation to reinforcement alignment.

Much of the versatility of composites may be attributed to the ability of the processor to orient reinforcements and provide directional properties. In the T joint shown in Figure 3-14, the advantages of orienting the fibers in (biased ±45°) fabric and woven fabric are illustrated.

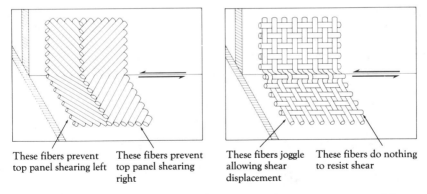

| These fibers prevent top panel shearing left | These fibers prevent top panel shearing right | These fibers joggle allowing shear displacement | These fibers do nothing to resist shear |

Figure 3-14. Comparison of ±45° biased fabrics versus woven fabrics. (HiTech Composites)

To overcome the unidirectional orientation or alignment problems, laminates or sandwich-type composites are constructed. These are similar to plywood. Each layer may be aligned or have fiber distributed in different directions. (See Laminates, Laminar.)

Laminar (Laminates)

Laminar composites are composed of two or more layers of materials bonded together. They are actually a class of composites and among the oldest known. Because the individual layers act as reinforcing and/or filler agents, they may be classed as a form of additive. These laminar layers are more than just a processing technique. (See Laminating.) Their selection, alignment, and composition constitute the performance properties of the laminar composite. The basic structural element of polymeric composites is the **lamina**. Lamina may be unidirectional fibers, woven fabrics, or sheets of material.

There are many possible combinations of laminar construction. Layers of the laminate may be isotropic, of the same material, or of different materials. They may be anisotropic with different layers aligned or oriented parallel or at various angles to each other. Each layer may perform a separate and distinct function. As a rule, laminar-reinforced composites are anisotropic. If all the materials are placed parallel to each other (0° lay-up), we think of the material as

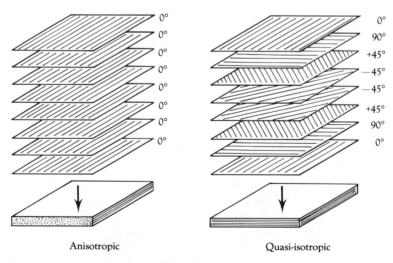

Anisotropic   Quasi-isotropic

Figure 3-15. Laminate plies tailored to meet specific requirements of composite design.

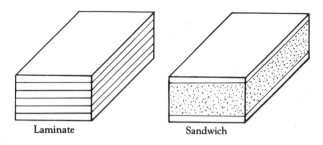

Figure 3-16. Two distinct classes of laminar composites.

being isotropic. Material with 0° lay-up on mandrels or in tank construction results in a composite with anisotropic properties. (See Figure 3-15.)

The composition of laminar-reinforced composites may be divided into two distinct classes: (1) laminates and (2) sandwiches. These are shown in Figure 3-16.

Laminates

Most materials (except wood) are isotropic and must be made artificially anisotropic in strength by fabrication of ribs, corrugations, or layers of selectively oriented materials.

Plywood is commonly cited as an example of a laminar composite composed of wood (anisotropic) reinforcing layers and a polymer matrix binder. Laminates are often prized for characteristics other than superior strength. They may greatly reduce material and labor costs, facilitate fabrication, reduce thickness, improve surface appearance, or overcome size limitations.

For the purpose of this text, only those laminar combinations containing organic materials will be included. **Metal-to-metal** laminates are important. The most familiar application is as clad United States coins. Clad metal to inorganic laminates are also important for applications in severe environments. These products should not be confused with ceramic-coated metal products which are not laminates. There are a few special applications where ceramic or different glass laminates are fabricated. **Inorganic-to-inorganic** laminates may be used as protective shields or for special lenses.

It is beyond the scope of this text to discuss matrices of glass, ceramic, or metal.

The combinations for organic laminar and matrices are variable and complex. This makes the study of laminar additives and composites one of the most challenging and potentially diverse options for the future. The idea of using different composites as layers or mixing sandwich and laminate layers further compounds the diversity of possible performance properties.

The idea of combining metal or inorganic plies with organic matrices is familiar. Safety glass is composed of plies of glass and polymer. Metal has been laid down in layers or plies with paper, cloth, or plastics to form selectively conductive layers for circuit boards. Felted metals (spongelike fibrous sheets), metal wood, wire, fibers, and mesh are used as laminates for protective housings, security, or radiation shielding, and as decorative construction panels. Many of the metal and inorganic laminates are fabricated by compressing layers of impregnated paper or cloth and the inorganic materials together under heat and pressure. (See High-Pressure Laminating.)

Organic-to-organic laminates composed of resin-impregnated paper (nonwoven) or cloth are used extensively. Plywood is an example of an organic-to-organic laminate. Decorative and industrial laminates are commonly used for counter tops, furniture facing, and in industrial applications as laminated gears, bushing, and dielectrics. Variations of compression molding processes are used to form these materials. Resin systems and organic fibers may be formed by hand lay-up methods. Aramid, carbon, graphite, and some hybrid reinforcing fiber cloths or filaments are made into a number of structural composites by hand lay-up, vacuum bag, and other techniques. Most use an epoxy matrix with various layers or plies of reinforcement.

For nonstructural applications, multilayered films and sheets are finding a growing market. Retortable, microwavable, seven-layer sheets are now being used to package a number of food products. The sheets are thermoformed to the desired container configuration. The sheet consists of an outer copolymer PP/adhesive/PVDC/adhesive/copolymer PP/homopolymer PP/clear homopolymer PP. The combination has the necessary performance properties needed for packaging food. It is aesthetically attractive, tough, shelf-stable, food safe, heatable, and possesses excellent barrier properties.

Some coextruded films are simply laminated to each other. Redi-Serve-Pak pouches of PVDC-coated, biaxially oriented polyamide have been laminated to a coextrusion of EVOH and LDPE. The pouches are used for tomato products, ice cream toppings, jellies, and other sauces.

An opaque sheet for high-temperature applications in aircraft interiors is composed of three layers of polyphthalate carbonate between skin layers of polyetherimide.

The technology and potential of multilayered films and sheets was one of the most significant developments of the 1970s. Although the concept seems obvious, so did the RIM processing techniques which are of equal importance.

Organic-to-inorganic generally brings the idea of an organic polyester or epoxy matrix and an inorganic glass lamina to mind. These composite laminates usually consist of several layers of glass fabric, filament, or mat bonded by a polymer matrix. Boat hulls, automobile bodies, large containers, pressure vessels, pipes, panels, and numerous other products are made of these materials. They are fabricated by a number of processes including matched-die, hand lay-up, filament winding, and extrusion.

Sandwiches

The ASTM has defined a sandwich as a combination of alternating dissimilar simple or composite materials, assembled and intimately fixed in relation to each other so as to use the properties of each to specify structural advantages for the whole assembly.

During World War II, the all-wood aircraft called the Mosquito was made of a sandwich composite. Layers of plywood (laminates) were assembled around thick cores of balsa wood. This sandwich construction was light, strong, stiff, and inexpensive. Both (laminates and sandwich) laminar composites depend on the interface adhesion. The relative surface energies of the two constituents, roughness, or coupling agents are of extreme importance. Differences between the expansion of dissimilar materials may result in locked-in stresses.

There are a number of possible core and facing materials for sandwiches. Various cellular, honeycomb, waffle, or corrugated materials are used as cores. (See Figure 3-17.) Properties of several foam (cellular) materials used as cores are shown in Table 3-8. Plastics, paper, metals, and ceramics may be made into all these core configurations. The skeletal structures formed by the core may be filled

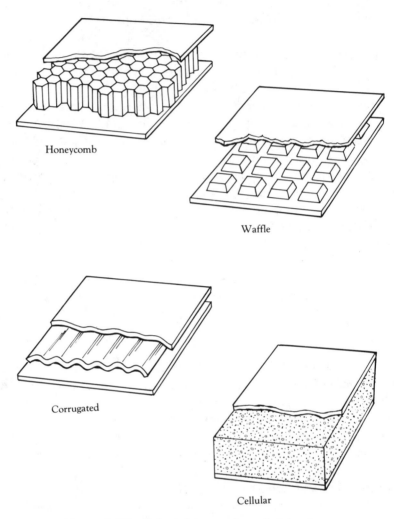

Honeycomb

Waffle

Corrugated

Cellular

Figure 3-17. Selected types of sandwich core.

with selected fillers. Ceramic and metal fillers are used in honeycombs to produce high-temperature, radiation-shielding materials.

As a rule, honeycomb, waffle, and cellular sandwiches are isotropic, with excellent thermal, acoustic, and strength-to-weight ratios. The corrugated core sandwiches are more anisotropic.

A number of thermoplastics are coextruded or combined with other films, metal foils, papers, or fibers to produce a number of

Table 3-8. Properties of Several Foam Materials Used as Cores

Type	Density (kg/m³)	Tensile Strength (MPa)	Compression Strength (MPa)	Maximum Service Temperature (°C)
Epoxies, rigid closed cell	80	0.35	0.62	177
Phenolics, foam-in-place	5–24	0.021–0.12	0.014–0.10	0.21–0.28
Polyurethane, rigid closed cell	21–48	0.10–0.65	0.10–0.41	82–121
Polyvinyl chloride, rigid closed cell	48	0.68–18.6	6.90	—

unique products. Some are true laminating processes and others may be classed as coating processes.

Aluminum foil and paper/aluminum foil are laminated to foamed boards for use as construction panels. A number of metal foils with outer layers of selected barrier or sealable polymers are used for packaging meats, jellies, nuts, and other food products.

Chapter 4.
How Are Composites Formed?

Introduction

The plastics industry may be divided into three large categories that sometimes overlap: (1) the material manufacturer who produces the basic plastics resins from chemical sources; (2) the processor who converts the basic plastics into solid shapes; and (3) the fabricator and the finisher who further fashion and decorate the plastics.

Technologists, engineers, and polymer chemists have been working for years to supply polymeric materials that were improvements of natural ones. The idea of improving or developing new polymeric materials accelerated rapidly after the turn of the century. The period from 1940 to 1970 has been referred to as the *plastics age*. There is every reason to believe that this plastics age will continue to outpace other industries.

The polymer industry is dependent on petroleum and other feedstocks. A flow chart of the polymer industry is shown in Figure 4-1. A study of global prospects for energy up to the year 2000 concludes that "oil is essential for just two major uses: transportation and petrochemical feedstocks."

This chapter presents a brief overview of processing techniques which form the vast number of polymeric products. (See Table 4-1.) The basic processes of the industry include molding, casting, thermoforming, expanding, coating, decorating, machining, finishing, assembly or fabrication, and radiation processing. The technology and proprietary information on each process have been cited in many volumes. (See Bibliography.)

126

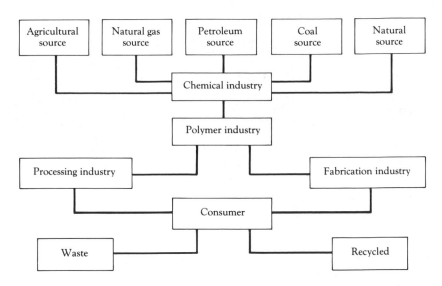

Figure 4-1. Flow chart of the polymer industry.

Molding Processes

For a process to be classified as a molding process, force (other than gravity or atmospheric) is used to help shape or mold the product. The broad use of the term **molding** means to shape on or in a

Table 4-1. Composite Product
and Percent Processing Methodology

Product Type	Hand Lay-Up	Filament Winding	Press Molding	Pul-trusion	Spray-Up	Auto-clave Molding	Other
Aircraft/ aerospace	10	10	25	5	1	40	9
Appliances	5	5	55	1	25	5	5
Construction	20	5	40	5	16	8	6
Consumer goods	10	1	50	1	40	—	—
Corrosion-resistant products	5	65	5	10	5	—	10
Electrical products	14	5	30	40	5	1	5
Marine	34	1	10	1	44	5	5
Transportation	14	5	45	5	20	1	10
Miscellaneous	25	5	20	5	25	5	15

mold, normally with heat and pressure. Molding processes include injection molding, compression molding, transfer molding, extrusion molding, blow molding, calendering, laminating, reinforcing, cold molding, cold stamping, sintering, and liquid resin molding.

Keep in mind that molding processes are used to form composite products. In some molding techniques such as extrusion, calendering, laminating, and reinforcing, the reinforcing agents may be continuous or very long. Other processes use short and mostly randomly oriented fibers, crystals, and fillers in the polymer matrix.

Injection Molding

This molding technique is similar to metal die casting. Many of the early machines were patterned after die casting machines. In fact, injection molded composites compete directly with die cast metals for many applications.

Thermoplastic and thermosetting materials may be injection molded. (See Polymer name.) The basic concept is to make the matrix molten and force it into a mold. When using thermoplastic materials, the mold is generally cooled to speed production rates. The part is held under the pressure of the closed mold until the material returns to the solid state. When thermosetting compounds are molded, the molten material is forced into heated molds where the part is cured into an infusible, insoluble solid. Fabrication of elastomers is similar except that vulcanization (cross-linked bonding) takes place after achieving the desired shape.

Injection Molding Thermoplastics

The fillers, reinforcing agents, and other additives are generally compounded and made into granular (e.g., pellets and powders) form ready for molding. The granules are fed into the machine from a hopper system. There are two basic types of injection unit. In the plunger type, a plunger forces the material around a heated cylinder and spreader (torpedo). There is little mixing or breaking of reinforcement fibers. (See Figure 4-2.) The reciprocating screw injection system is most popular. In this system, the granules are forced forward by the screw action. The auger-type action generates thermal energy from the friction and shear as the material is forced forward. The selection of screw configuration depends on the material being processed. (See Extrusion.) The barrel is heated to ensure plasti-

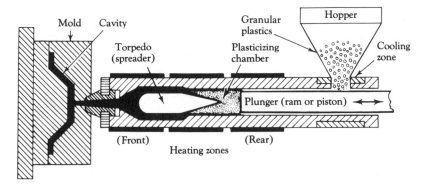

Figure 4-2. Simple schematic of plunger-type injection molding.

cation (i.e., softening to allow plastic deformation) and reduce molding cycle times. To prevent hot materials from being forced back into
the screw and from drooling while the mold is in the open position,
a nonreturn check valve is needed.

There are several machine technologies that help reduce the plas-
tication time, thus reducing the cycle time. More than one screw
may be used, or a preplasticator screw may be used to feed a plunger
cylinder. These designs reduce the **injection time** (i.e., the time it
takes to force the melt into the mold cavity). All cycle times must be
reduced if high production rates are to be maintained. The **dwell
time** is the pause in pressure which allows degassing before the
mold is closed. Sometimes the term **dwell** is used to describe the
time that pressure is kept on the material in the mold. The time
required for the material to cool in the mold is called **freeze time**.
Dead time is the time required to open the mold and remove the
part. The time the machine is not in operation is called the **down
time**.

Most machines are rated by the shot capacity expressed in ounces
of PS and plasticating rate in ounces per second.

All machines use one of three basic clamping units to open and
close the mold during the cycle time (injection, dwell, freeze, and
dead). **Mechanical** toggle clamps are the most popular because they
are fast and easy to adjust. **Hydraulic** clamps use hydraulic cylinders
to open and close the mold. **Hydromechanical** clamps vary greatly
in design and rely on hydraulic action to help develop the full clamping force.

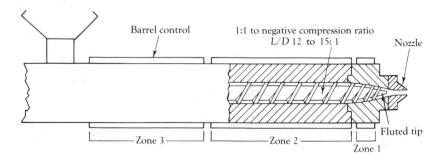

Figure 4-3. Thermoset screw barrel assembly.

Injection Molding Thermosets

There are several differences in the basic design operation when molding thermosetting compounds. It is important to prevent overheating and possible precure of materials in the machine barrel. A nonreturn check valve is not required in these machines since the material is very viscous and little material is left in the barrel after injection. The screw flights are shallow with 1:1 compression ratio on the screw. (See Figure 4-3.) With BMC or other heavily filled or reinforced materials, a plunger-type machine is used. Material temperatures in the barrel of either type are carefully controlled and maintained [130–240°F (54–116°C) depending on material]. When the carrier films are removed, SMC may be placed into the feeds of injection molding machines.

The mold is usually about 200°F (93°C) higher than the barrel nozzle. Insulating plates are generally used between the molds and clamping units to prevent excessive heat loss by conduction. The molds are generally heated by hot oil or electrical coils until the molded part is cured. Hot runner systems are used for both thermoplastic and thermoset materials to reduce runner and sprue scrap. (See Design Practice.)

Coinjection Molding

Coinjection molding, sometimes called sandwich molding, is a process where two or more materials are injected into the mold cavity. The composite parts may have a cellular core and solid skin surfaces. The layers of solid skin or foam may be of different but compatible materials. Figure 4-4 illustrates the basic conception of coinjection molding.

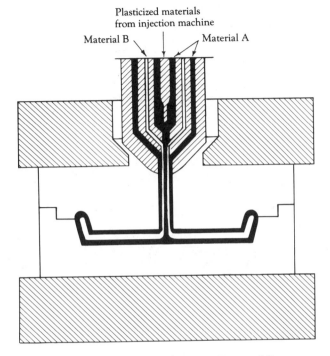

Figure 4-4. Concept of coinjection molding.

Several different machine configurations are available. Two-stage preplasticating systems and reciprocating screw systems are the most popular. Fibers stiffen the polymer matrix and flow patterns may result from orientation of the fibers. Light, less costly, core materials may be used for many applications. Only the exposed skin surfaces need to be plateable or pigmented. Products have little molded-in stresses. Business machine cases, outdoor furniture, and phonograph bases are familiar applications. (See Expansion Processes.)

Reaction Injection Molding

Reaction injection molding (RIM) is a process where two or more monomer components are forced into an impingement chamber. This mixture is then injected into a mold cavity. Rapid polymerization occurs from reactions among the components in the cavity. A cleaning piston is used to clean the mix chamber after each shot. (See Figure 4-5.) The relatively stress-free product may be produced from a number of materials and offers a range of densities. Products

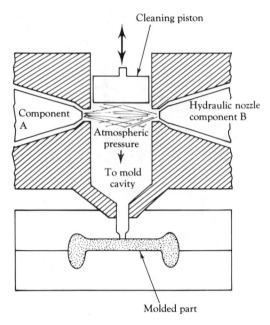

Figure 4-5. Schematic of RIM.

may be flexible or tough and rigid. The process has also been called liquid reaction molding or high-pressure impingement mixing. When short fibers or flake (particulates) are used to produce a more iso-tropic product, the process is called reinforced reaction injection molding (RRIM). Like other processing parameters, fiber loading increases monomer viscosities and abrasive wear on flow surfaces.

Since 1970, the most used polymer has been polyurethane. Poly-urethane/urea hybrid, epoxy, polyamide, poly(dicyclopentadiene), polyurea, and polyurethane/polyester hybrid systems are used. Many of these parts compete directly with coinjection and structural foam molded components. (See Expansion Processes.) Applications include business machine housings, rocker arm covers, bumper covers, exterior doors, quarter panels, fenders, and front and rear automobile fascia.

Compression Molding

Compression molding is one of the oldest known molding processes. The concept is analogous to making waffles. Generally, pre-heated, premeasured compounds are placed into an open mold. The mold is closed under heat and pressure until the part is fully formed and cured. Preforms or screw plasticators may feed automated systems. Both thermosetting and thermoplastic polymers are used. The process is popular for several reasons:

1. Heavily filled and reinforced compounds including BMC and SMC are used.
2. There is little orientation of resin or additives caused by the forming process.
3. Equipment and tooling costs are low because of the low pressures.
4. There is little material waste or postfinishing.

Most molding presses are hydraulic. The platens are heated by steam, electric, or oil, with electric being the most popular because it is fast, hot, and clean.

The principle of compression molding is shown in Figure 4-6. There are three basic mold designs:

1. Positive mold designs do not allow any material to escape.
2. A flash mold permits the escape of excess molding material. Flash is removed before the part is considered finished.
3. The most popular mold design is called semipositive. This design allows an amount of excess material to escape as it becomes fluid.

Because many thermosetting polymers degas upon cure, it is common to open the mold briefly to allow any trapped gases to escape. Products commonly produced by compression molding include

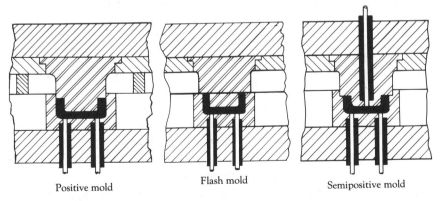

Positive mold   Flash mold   Semipositive mold

Figure 4-6. Three basic compression mold designs.

building panels, dinnerware, machine housing, and electrical components.

Calendering

Only a few composite materials are made by calendering. The calendering process consists of squeezing thermoplastic polymers between heated rollers. The gap (nip) between the last pair of heated rollers determines the film or sheet thickness. Reinforcing fabrics or other substrates may be fed into the third nip for bonding. It can be bonded to or between layers by passing the hot composite between other rollers. Extrusion and postprocessing rollers have displaced many calendering applications. Calendering equipment is costly and the rollers are crowned to compensate for deflection.

Blow Molding

Blow molding is a process used to produce hollow objects such as fuel tanks, milk containers, and some toys. The technique was adapted from the glass bottle industry. The basic concept is to inflate a hot hollow tube inside a closed mold. Once the polymer has cooled, the mold is opened and the part ejected. There are three basic techniques used to produce blow molded objects: The major difference is in the way that the hot hollow tube called the **parison** is formed.

Extrusion-Blow Molding

Extrusion-blow molding uses an extruder to melt the polymer and form the hollow tube. Composite, two- or three-layer parisons may be formed. If the extruder constantly melts and continuously forms the parison, the machine system is referred to as a **continuous** blow molder. If the melt is placed into an accumulator (holding chamber) before the parison is formed, the system is called an **intermittent** blow molder.

Continuous systems are used extensively for small containers. The parison extrusion rate is generally less than 1000 lb/h. A manifold, multihead system may be used to produce more than one parison. A wheel or shuttle press system may be used to move a number of molds past the parison for blowing. This system is used when high volume and production are required. Nearly 80% of the cycle time is used in cooling the object once it is formed. Ambient air may be used in blowing and cooling the inside of the part. Chilled air greatly reduces the cooling cycle. Even faster cooling cycle time may be obtained by using CO_2 gas. Not all parts or production needs warrant the speed or cost of this gas. With some polymers, a post-mold cooling cycle is sufficient. With this system, once the polymer has sufficiently set to retain its blown shape, it is removed to a postcooling station where cool air is circulated inside and outside the object.

Intermittent systems are used to form large parts. The major difference is that a quantity of material is held in an accumulator until the parison is to be formed. Once the large parison is formed, additional material is accumulated as the blow molding operation and cooling cycle are completed. This system may accumulate up to 300 lb with parison ejection rates of 20,000 lb/h.

Several accumulator designs are available. In one accumulator head system, the arrangement ensures that parts will not have a weld line even when using high-viscosity polymers. In this system the hot melt is forced to separate around a mandrel and recombine in the accumulator. The technique ensures that the two melt streams overlap as they recombine. The intermittent system then completes the parison and forming cycle.

In ram accumulator systems, the melt is fed into the accumulator before being forced out of the die head to form the parison. In reciprocating-screw systems, the melt is accumulated in front of the screw. The screw is then forced forward, causing the melt to be forced out of the die head and to form the parison.

Injection-Blow Molding

Injection-blow molding is a two-stage process where the parison is injection molded and then transferred to the blow station. There are two molds involved: one to shape the parison preform, and one to mold the part. There are several important advantages with this system:

1. Part wall thickness can be carefully controlled with close tolerances.
2. There is no waste from the pinch-off.
3. No postfinishing operations are normally required.
4. There are some physical and mechanical improvements because of controlled flow orientation.

5. There are no pinch-off or weld lines on the bottom of the container.
6. Coinjection molded parisons with different polymers may be used in composite designs.
7. Fiber and particulate reinforcements are more easily processed using this system.
8. Preforms may be made or purchased and stored for later production.
9. Large numbers of containers are currently products of PAN, PET, and PVC using this processing technique.

Stretch-Blow Molding

Stretch-blow molding is a process where the preform is stretched both axially and radially. This biaxial orientation improves strength, clarity, and drop impact. In single-stage systems, the preform is made and then stretched and blown in the mold. With two-stage systems, the preform may be manufactured and later stretched and blown into the desired shape on another machine. Many oriented PET containers are produced by stretch-blow molding. Some of the common thermoplastics that are stretch-blow molded are shown in Table 4-2.

Multilayer-Blow Molding

Multilayer-blow molding is used when several layers are needed to produce the desired composite performance properties. Carbonated beverage containers are produced by injection-blow molding in multilayer systems. Barrier layered PET containers are used for some soft drink containers. Coinjection-molded preforms may be used to produce the three-layered preform. The barrier layer is normally sandwiched between two skin layers. In some food containers as many as seven layers may be needed.

Table 4-2. Approximate Melt, Stretch Ratios, and Stretch Temperature of Selected Polymers

Material	Melt (0°C)	Stretch Ratios	Stretch (0°C)
PET	280	16:1	90–120
PVC	180	7:1	100–120
PAN	210	9:1	100–130
PP	240	6:1	130–150

Extrusion Molding

The extruder machine is an unusually versatile machine. It may be used in mixing BMC or in forcing the melt through a die or into a mold. The process is similar to metal extrusion.

The extrusion process plasticizes (melts and mixes) powdered or granular polymers into a (generally) continuous melt. The melt stream is then sent to the die, mold, or accumulator. (See Extrusion-Blow Molding.) Screw technology determines the output, milling rate, die pressure, and type of material that is to be plasticated. The screw rotates in the extruder barrel. The length (L) and barrel diameter (D) are expressed as an L/D ratio. A 40 in. length and 2 in. bore diameter would be expressed as a 20:1 L/D extruder. A cross section of a typical screw extruder is shown in Figure 4-7. Extruders are sold by barrel size and the amount of material that they plasticate per minute or hour. Commercial extruder bores may vary in size from less than 2 in. to more than 12 in. The amount of LDPE that they extrude varies from less than 5 lb/h to more than 11,000 lb/h.

For many operations, multiple-screw extruders are used, for example, in the compounding and extrusion of many reinforced thermoplastic and thermosetting composite materials. Since thermosets are inherently heat sensitive, shorter L/D screws (12:1) are used to reduce the amount of heated materials stored in the screw flights and the amount of time in the barrel. The compression section of the screw usually has a compression ratio (back diameter to front) of 1:1. With twin-screw machines, it is possible to introduce con-

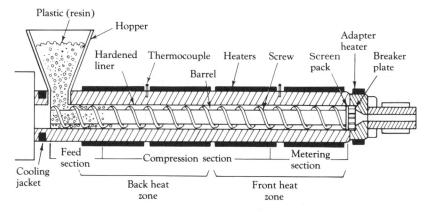

Figure 4-7. Simple schematic of extruder.

tinuous strands of reinforcement at the hopper. The screw will break the fibers into short lengths. Adding fibers to a premelted polymer will reduce breakage and extruder wear, while increasing dispersion and fiber wet out.

Most thermoplastic extruders use a screen pack held by a breaker plate to filter out foreign material and create back pressure on the polymer. When these screens become clogged, they must be changed or cleaned. Slide plate and continuous flow screen changers are commonly used. With thermosetting materials it is imperative that the die head be easily removed to purge in full any unwanted or semicured material. When using reinforcing agents, screws and barrels must be hardened or specially coated for added wear resistance. Most wear occurs at the tip and metering section. Barrel inserts or liners are used to replace or repair worn bores.

A number of extruder techniques are available. The most important are for compounding, producing profile shapes, pipe, fibers, films, and sheets. Extrusion coating and coextrusion are further variations.

Wire and **cable** extrusion is actually a coating operation. (See Coating Processes.) Wires, cables, metal strips, rope, and other profile shapes may be coated by the hot melt as it emerges from the crosshead die.

The distinction between extrusion coating of paper, fabric, cardboard, foils, or other substrates and laminating of various layers is not clear. There is no clear direction in the literature where some of these techniques are to be placed. In this text, all coating operations (including extrusion) are discussed as coating processes. Operations where several layers of materials are placed together to form a composite product will be described as a laminate. (See Laminating.) Structural foams and extruded forms with integral skins are described as expansion processes.

This problem is further exacerbated by newer techniques that combine several operations as required in the production of some composites. In one operation a coextruded film with a cellular sandwich core is coated with a solvent cast layer on the top side while a continuous spray-up operation is accomplished on the bottom surface. This "laminate," a preform, or SMC is then formed by a pressure-bag molding technique.

Extruders are used to premix and compound resin, filler, reinforcements, and other additives. Some types of BMC and specially shaped premixes, including granular compounds, are made by the extruder. Many of these compounds are formed by other molding operations.

Many different **profile** shapes may be made by forcing the melt out of the die orifice. The orifice must be carefully designed and shaped because the extruded polymer will normally swell as it leaves the die. In Figure 4-8 the relation between die orifice and extruded profile shape is shown. Profile shapes may be postformed or sized. Flat ribbon shapes may be passed through sizing plates or rollers to form an L-shaped profile.

Pipe or tubular forms are extruded over a pin or mandrel. The extrudate is then sized and quenched in a water cooling trough. These products may be made any practical length. Tube diameter and polymer orientation may be partially controlled by the take-up mechanism.

Fibers may also be extruded from the die and drawn onto the take-up or bobbin. The openings through which the fibers are shaped are sometimes called spinnerets (after the opening under the jaw of the silkworm). Forming fibers through these spinneret openings is referred to as spinning.

Not all polymers may be processed by melt spinning. Some must be dissolved in a chemical solvent and then passed through the spinneret into a coagulating bath. This bath makes the plastic be-

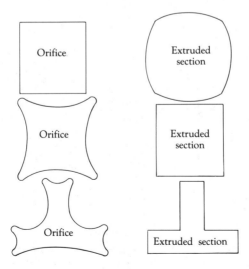

Figure 4-8. Relation between die orifices and extruded sections.

come a solid filament. The process is called wet spinning. In solvent or dry spinning, a solvent is used to soften the polymer. Once the solution is forced out of the spinneret, the solvents are evaporated from the fibers.

Film and **sheet** extrusion are accomplished by several different techniques.

There are two basic methods of producing films (less than 0.01 in. in thickness). One important method is blow extrusion (lay-flat or tubing). As the melt emerges in a tube form, air is forced onto the tube causing it to form a bubble. Once the polymer has cooled, it may be rolled up on spools as seamless film for packaging of some foods or garments. One side of the film may be cut during take-up. An emerging tube with a diameter of 1 ft may be blown to three times the diameter. If it is slit and opened, it will be nearly 10 ft wide. Blow extrusion films may be oriented in several ways. Orientation results in blowing and stretching the tube by take-up pressure. The actual size and thickness of the finished film is controlled by extrusion speed, take-up speed, die (orifice) opening, material temperature, and air pressure in the bubble.

Films and sheets are formed by extruding materials through dies with a long horizontal slot. The die lands or nip (pinch) rollers may further shape or control the thickness of the film. The takeoff equipment and tentering chains may be used to help orient film and sheet materials longitudinally and transversely. (See Figure 4-9.)

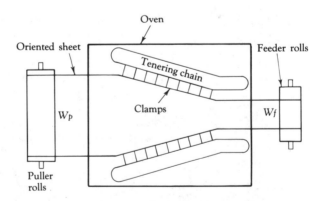

Figure 4-9. Film and sheet are oriented both longitudinally and transversely. The ratio of feeder-roll to puller-roll speed is 1:4. The ratio of width W_p to width W_f is 3:1.

The term **sheeting** is used to describe flat pieces of sheet (greater than 0.010 in. in thickness) polymer which have been cut to convenient dimensional sizes. Sizes are extruded, calendered, and cast. (See Casting.)

Laminating

Laminating is a molding process of producing a laminated product. Laminated, laminate, and laminated molding are terms used to describe a composite form or product made by the process of laminating.

Simply stated, laminating is the process of combining two or more different layers into one composite piece. These may be superimposed layers of filler, reinforcement, polymers, impregnated or coated paper, or other materials. Examples may include metal and polymer laminated food packages, honeycomb structural composites, or decorative laminates.

Sometimes it is confusing or difficult to distinguish between laminating and reinforcing processing. (See Reinforcing.) In this text, laminating will refer only to processes or laminated forms where there are separate phases or two different types of material being combined in layers. Mechanical or chemical adhesion may be used to bond the layers together. In many laminated forms, mechanical and chemical bonds are used (e.g., metal to plastics to plastics).

Laminating includes the following techniques: (1) high-pressure, (2) continuous, and (3) extruded sheet and film laminating techniques.

The term **high-pressure laminate** refers to a laminar composite that is in a rigid sheet, rod, or tube form. The word "high" pressure indicates that forces greater than 1000 psi were used. The terms low-pressure laminate and high-pressure laminate are used in the literature to make a distinction between rigid laminates and those made into other shapes by reinforcing techniques. In this text, the term high-pressure laminating is used to describe a technique of bonding resin-impregnated materials together by heat and pressure. A number of decorative and industrial laminates are made by this method. Resin or polymeric matrices such as PF, DAP, and MF may be used with layers of cloth, paper, or foils to produce the composite laminar.

A typical arrangement of layers in a decorative high-pressure lam-

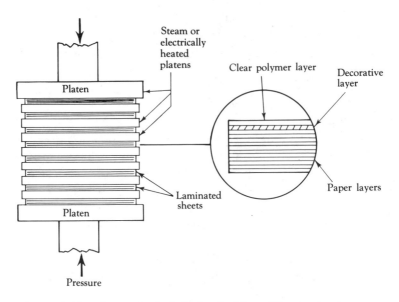

Figure 4-10. Concept of multiple stacking of laminates in press.

inate is shown in Figure 4-10. A number of laminates may be produced at one time. Most decorative laminates are cemented to a wood
backing in the manufacture of kitchen cabinets, furniture, and table
tops. Industrial laminates are used extensively as printed circuit
boards and electrical insulating materials or machined into gears
and bearings.

Composite sheet materials may be made by **continuous laminating** techniques. In continuous lamination, fabrics, paper, or other
reinforcing materials are impregnated with partially polymerized
B-stage resins. Release films are then placed on the two surfaces as
the composite is heated and pressed to form the finished laminate.
The concept is illustrated in Figure 4-11.

In a similar process, films may be combined with other films,
foils, paper, or fabrics to produce a laminated product. Thermal
laminating machines use heat and pressure to bond the materials
together. Dry adhesives may also be used and activated by heat and
pressure. In wet bond laminating the solution adhesive is commonly
cured by ultraviolet or ionizing radiation. Some of these techniques
may be used in flexible printed circuit boards, as packaging materials, and as drapes for welding or solar shields.

Extruded sheet and **film laminates** could be included as a con-

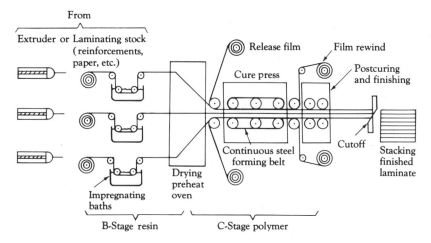

Figure 4-11. Continuous lamination from extruders or laminating stock from B-stage resin to infusible C-stage polymer.

tinuous process. A number of polymers may be extruded into a true composite laminate. The layers may be different materials, colors, or compositions. These products are used as engraving stock, re-frigerator liners, construction panels, and stock for thermoforming containers. The concept of a composite laminate is illustrated in Figure 4-12. Reinforcements or other core materials may be fed into the melt and pressed into the composite laminate stock. Many lam-inates are postformed by thermoforming or stamping in matched dies. (See Cold Stamping.)

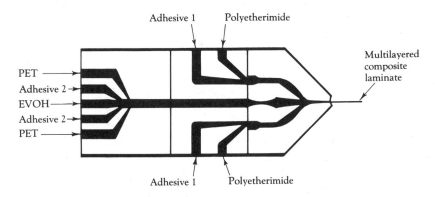

Figure 4-12. Continuous extrusion of multilayered laminate.

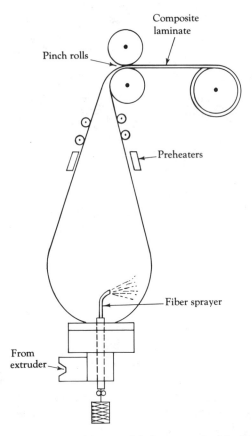

Figure 4-13. Composite blow molded material with fibrous layer.

In one proprietary process, composite sheets are produced by filament reinforcing the inside of the hot blown film and pressing the fibrous layered film between pinch rollers. This concept is illustrated in Figure 4-13.

Sandwich construction consists of two relatively dense, thin, strong outer layers or skins separated by a lightweight core material. This unique composite is sometimes called a structural sandwich because it is used in many structural applications. These lightweight, load-carrying materials are used in the manufacture of cargo containers, portable shelters, architectural walls, skylight systems, space vehicles, airliners, helicopters, navy ship interiors, snow skis, doors, cabinets, and radomes.

The surface skins of a sandwich composite must be strong. Like the outer flanges of any I-beam, they must also carry the axial and in-plane shear loading. Most of the tensile and compressive forces are exerted on the upper and lower skins. The facing materials may be composed of aluminum, titanium, stainless steel, polymers, wood, impregnated paper, fibrous glass, graphite, carbon or other materials.

The core material may be solid, cellular, honeycomb, or any other configuration. It must transmit the loads from one facing to the other. There is little penalty in the performance when a solid core is replaced with less-dense materials. Solid wooden cores have commonly been used in doors, snow skis, boat hulls, pallets, and furniture. Balsa wood was used as the core between thin layers of plywood in the sandwich construction of the all-wood Mosquito aircraft of World War II. Solid wood and plywood are relatively low-cost core materials used today in the sandwich construction of large shipping containers. Foam or cellular plastics are commonly used as core materials in sandwich construction. The adhesive attachment of the outer skins has been made with polystyrene, polyurethanes, and polyvinyl chloride foams. Thermal and electrical conductivities of the sandwich composite depend on the selection of facings, core, and bonding agent. Various facing materials are used on foamed cores. Foamed sandwich panels are used in the manufacture of refrigerator liners in truck boxes, railcars, food coolers, and exterior panels for mobile homes.

A sandwich composite may be produced by a process known as foam reservoir molding or elastic reservoir molding. An open-cell polyurethane foam is impregnated with resin (epoxies) and placed between two skin layers. The composite is then squeezed. This forces some of the resin to come into contact with the skin layers. The foam and matrix become a catacomblike skeletal structure. When cured, the strong, lightweight sandwich composite is removed. Applications may include doors, hoods on vehicles, and construction panels.

Honeycomb core materials are among the strongest possible structures for a mass with adjacent cells. Resin-impregnated kraft paper, aluminum aramid paper, glass-reinforced polymers, titanium, stainless steel, and other materials may be made into various honeycomb configurations.

The physical and mechanical properties of the honeycomb are dependent on the basic composition, cell size, and cell geometry. All honeycombs are anisotropic. Large cells or overexpansion on the honeycomb pattern will lower density and change the strength level of the core-to-face attachment. If cell walls are too thin, the bend and shear strengths of the sandwich will be lower. Two major methods of producing honeycomb core materials are illustrated in Figure 4-14.

Phenolic resin-impregnated paper is another low-cost core material. It is widely used in the manufacture of doors, walls, kitchen cabinets, and containers. Many of the cell configurations are custom designed. Corrugated and fibrous reinforced paper are used. Aramid paper honeycomb core materials are used in sandwich panels for the interiors of aircraft. Selected properties of paper honeycombs are shown in Table 4-3.

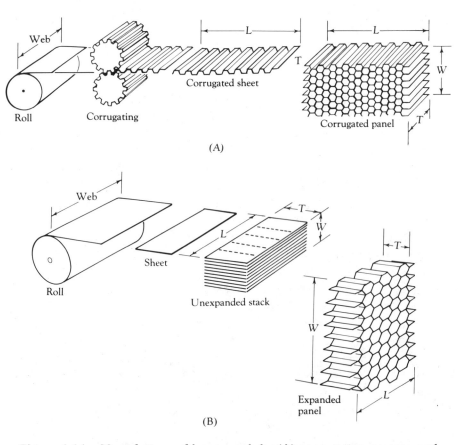

Figure 4-14. Manufacture of honeycomb by (A) corrugation process and (B) expansion process.

Table 4-3. Properties of Hexagonal Nomex Paper Honeycomb

| Honeycomb Designation Material–Cell–Density Gauge | Compressive | | | | Plate Shear | | | |
| | Bare | Stabilized | | L Direction | | W Direction | | |
	Strength (kPa)	Strength (kPa)	Modulus (MPa)	Strength (kPa)	Modulus (MPa)	Strength (kPa)	Modulus (MPa)
HRH 10–1/8–3.0(2)	2,068	2,275	137	1,241	48	655	24
HRH 10–1/8–4.0(2)	34,473	3,861	193	1,689	63	965	32
HRH 10–5/32–5.0(4)	5,515p[a]	6,205p	—	2,482p	72	1,241p	34
HRH 10–3/16–2.0(2)	1,034	1,172	75	758	28	379	15
HRH 10–3/16–4.0(2)	3,447	3,861	193	1,689	53	965	32
HRH 10–1/4–2.90(2)	1,034	1,172	75	758	28	379	19
HRH 10–1/4–4.0(5)	2,551	2,757	—	1,654	51	861	24

[a]p = preliminary property.

Table 4-4. Properties of 5056, 5052, and 2024 Hexagonal Aluminum Honeycomb

Honeycomb Cell–Material–Gauge	Nominal Density (kg/m³)	Compressive			Plate Shear			
		Bare Strength (kPa)	Stabilized Strength (kPa)	Stabilized Modulus (MPa)	L Direction Strength (kPa)	L Direction Modulus (MPa)	W Direction Strength (kPa)	W Direction Modulus (MPa)
5056 Hexagonal Aluminum Honeycomb								
1/16–5056–0.0007	101	6,894	7,584	2,275	4,447	655	2,551	262
1/16–5056–0.001	144	11,721	12,410	3,447	6,756	758	4,136	344
1/8–5056–0.0007	50	2,344	2,482	668	1,723	310	1,068	137
1/8–5056–0.001	72	4,343	4,619	1,275	2,930	482	1,758	262
5/32–5056–0.001	61	3,275	3,447	965	2,310	393	1,413	165
3/16–5056–0.001	50	2,344	2,482	669	1,758	310	1,069	138
1/4–5056–0.001	37	1,413	1,448	400	1,172	221	724	103
5052 Hexagonal Aluminum Honeycomb								
1/16–5052–0.0007	101	5,998	6,274	1,896	3,516	621	2,206	276
1/8–5052–0.0007	50	1,862	1,999	517	1,448	310	896	152
1/8–5052–0.001	72	3,585	3,758	1,034	2,344	483	1,517	214
5/32–5052–0.0007	42	1,379	1,482	379	1,138	255	689	131
5/32–5052–0.001	61	2,723	2,827	758	1,862	386	1,207	182
3/16–5052–0.001	50	1,862	1,999	517	1,448	310	896	152
3/16–5052–0.002	91	5,309	5,585	1,517	3,172	621	2,068	265
1/4–5052–0.001	37	1,138	1,207	310	965	221	586	112
1/4–5052–0.004	127	9,377	9,791	2,344	4,826	896	3,034	364
3/8–5052–0.001	26	586	655	138	586	145	345	76
2024 Hexagonal Aluminum Honeycomb								
1/8–2024–0.002	107	7,584	8,446	2,068	5,240	814	3,241	310
3/16–2024–0.0015	56	2,275	2,551	593	1,999	379	1,241	159
1/4–2024–0.0015	45	1,517	1,724	276	1,379	290	827	131

Metal honeycombs are relatively expensive. Aluminum and aluminum alloys are the most commonly used honeycomb core materials. Aluminum honeycomb core materials are available in a number of alloys, cell shapes, and foil gauges. Selected properties of hexagonal aluminum honeycombs are shown in Table 4-4.

Aluminum I-beams are used as the core materials for some translucent sandwich panels. These beams are welded into a structural frame with a decorative grid pattern. Translucent reinforced skins are then bonded to each side to produce a translucent sandwich for architectural walls or skylight systems.

A number of fibrous-reinforced polymer honeycomb materials are available. They may be manufactured with a variety of fiber and polymer formulations. Polyamide, phenolic, silicone, and polyesters are commonly used. Comb stock materials may be made by compression, pultrusion, or matched-die techniques. Selected properties of glass-reinforced plastic honeycombs are shown in Table 4-5.

The adhesive attachment is critical in transmitting shear and axial loads to and from the core. Liquid resins, supported film adhesives, and unsupported and reticulating films are used to bond

Table 4-5. Properties of Several Glass-Reinforced Plastic Honeycombs

Honeycomb Material–Cell–Density	Compressive			Plate Shear			
	Bare	Stabilized		L Direction		W Direction	
	Strength (kPa)	Strength (kPa)	Modulus (MPa)	Strength (kPa)	Modulus (MPa)	Strength (kPa)	Modulus (MPa)
Glass-Reinforced Polyimide Honeycomb							
HRH 327–3/16–4.0		3,033	344	1,930	199	896	68
HRH 327–3/16–6.0		5,377	599	3,171	310	1,585	103
HRH 327–3/8 –4.0		3,033	344	1,930	199	1,034	82
Glass-Reinforced Phenolic Honeycomb (Bias Weave Reinforcement)							
HFT–1/8 –4.0	2,688	3,964	310	2,068	220	1,034	82
HFT–1/8 –8.0	9,997	11,203	689	3,964	331	2,344	172
HFT–3/16–3.0	1,896	2,585	220	1,378	165	689	62
Glass-Reinforced Polyester Honeycomb							
HRP–3/16–4.0	3,447	4,137	393	1,793	79	965	34
HRP–3/16–8.0	9,653	11,032	1,131	4,551	234	2,758	103
HRP–1/4 –4.5	4,344	4,826	483	2,068	97	1,172	41
HRP–1/4 –6.5	7,067	8,136	827	3,103	172	1,793	76
HRP–3/8 –4.5	4,205	4,757	448	2,068	97	1,172	41
HRP–3/8 –6.0	6,205	6,895	689	2,758	155	1,793	69

the layers' components together. Polyimide, epoxy, and phenolics are used to bond core pieces together. Some metals are resistance welded. Resin-impregnated fiber matting, cloth, or paper may be used in the core-to-face bond. Unsupported and reticulating films contain only the adhesive, which is activated by chemical or thermal activity. The major objective of any adhesive in honeycomb construction is to have the adhesive adhere on the cell edge and to have as large a fillet as possible. Some manufacturers place adhesives only on the cell edges. In some sandwich construction, the foam core forms the adhesive attachment as the polymer expands against the skin layers.

The manufacture of sandwich composites varies in technique from simple hand lay-up operations to vacuum bag, autoclave techniques. Some panels are manufactured on continuous sheet production lines as illustrated in Figure 4-15. SMC and other skin materials may be used in press laminating in matched-die operations requiring ribs or more complicated shapes.

Reinforcing Processes

The term **reinforcing** implies that an agent has been added to improve or "reinforce" the product. For many years the term **reinforced** simply meant any polymer made stronger by the addition of glass or asbestos fibers. Today, many reinforcements are used and the technology once employed has changed dramatically.

In 1940, the term **reinforced plastics** meant any process using low pressure or no pressure to form a fibrous-reinforced product. The term high pressure was reserved for those processes used to form melamine and/or phenolic-impregnated paper or fabrics under high heat and pressure.

Today, we generally class all reinforcing processes together. There have been attempts to codify these processing techniques but none

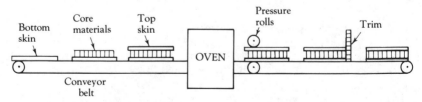

Figure 4-15. Continuous sheet production of sandwich composites.

seems satisfactory. Reinforcing processes have been divided into low–high pressure techniques, open–closed molding techniques, low–high temperature techniques, and laminating–reinforcing techniques. Technological advances continue to present new exceptions. The terms **advanced, high-strength,** or **structural composites** began to be used in the 1960s to imply that a new material and processing technique had evolved. It implied that a stiffer, higher-modulus material with exotic reinforcements in new matrices was being used. Reinforced plastics/composite (RP/C) materials have evolved into an important industry.

In this text the term **reinforcing** is used to describe a number of processing techniques. Reinforced plastics (RP) does not seem descriptive enough. All polymers may be reinforced and all processes may be used to form reinforced plastics products.

Reinforcing processes may include the following: (1) matched-die molding, (2) hand lay-up, (3) vacuum-bag molding, (4) pressure-bag molding, (5) spray-up molding, (6) filament winding, (7) centrifugal reinforcing, (8) pultrusion, (9) continuous reinforcing, and (10) cold-mold techniques thermoforming. Most of these processes may be classed as laminating techniques by definition if two or more different materials are combined in layers. Rigidized-shell techniques are variations of spray-up but could be more accurately classed as a laminating method.

Matched-Die Molding

Matched-die molding employs two major parts referred to as male and female dies. These matched dies are used to make molding compounds conform to the space left between the two parts. High-quality, precision composites may be made by matched-die molding. Most operations use preforms to lower costs and become competitive. Reinforcing mat preforms, BMC, SMC, TMC, and XMC are used in automated matched-die systems. Dies are heated to control cure with pressures under 1000 psi for softer BMC.

There are three basic materials used in matched-die molding: (1) mat or preform, (2) bulk molding compounds, and (3) sheet molding compounds. (See Reinforcements, Fillers, Additives.)

Early matched-die techniques used hand lay-up reinforcing and compression molding machines to control thickness and produce a finished surface on both sides of the part. In some operations fibrous materials are randomly chopped and made to conform to the general shape of the desired part. The matrix binder may be preimpregnated into the fibers or accomplished during the mold loading. The formulation of matrix and fibrous reinforcements at the press is impractical except for short-run or special-design products.

Bulk molding compounds (sometimes called premix or dough molding compounds—DMC) offer some advantages over preform processing. BMC is formulated with all additives into a fibrous putty form. It may be extruded or shaped into convenient, carefully measured charges. They may be made into H-beam or other profile shapes and automatically fed into the matched die. The chemical thickening agents, thermoplastic additives, and variability of resin, filler, and reinforcement types permit dependable tailoring for a specific design or performance requirement. BMC are used in compression, transfer, extrusion, and injection molding operations.

A major market for BMC parts is the transportation and appliance industry. Shower floors, heater housings, and appliance cases are typical examples. The products are quickly molded, light, strong, and resist deterioration in the intended environment.

The idea of combining additives and resin into a convenient sheet form has replaced many of the preform and mat techniques. Sheet molding compound is used in matched-die systems. It comes in a thin, semitacky sheet and is cut to the desired size and plied to suit design and performance needs. Both BMC and SMC may be molded in complex shapes with ribs, bosses, and cutouts. BMC are isotropic with fiber lengths generally less than 0.025 in. Fiber content—not length—greatly influences part strength of BMC. A wider range of fiber lengths may be used in SMC. If long, continuous fibers are used, the product will have great strength in the direction of the fiber. Although most SMC are formulated with fibers varying in length from 0.025 to 1.00 in., long fiber plies may be placed in the mold to increase strength in a specific direction. The name SMC also implies that large area parts are produced. There are a number of automotive and aerospace applications. Body panels, truck cab components, hoods, small boat hulls, furniture, and appliance components are familiar products.

There are several designations and variations of SMC devised to meet a diversity of design parameters. In some designs requiring thick sections and heavy rib details, thick molding compounds (TMC) may be used. TMC are available in sheets from up to 2 in. thick.

(See Sheet Molding Compounds.) Another SMC variation designed to provide high strength in the dominant fiber direction is the XMC. This preform compound uses continuous fibers arranged in an X pattern. Resin-impregnated fibers are wound at various angles on a mandrel. A protective film is then wrapped over the final layer and the preform is removed. The XMC is then cut and flattened for storage and maturation.

Hand Lay-Up

Because no pressure (other than atmospheric) is used to form the composite shape, this reinforcing process should not be called a molding technique. Contact molding or open molding are more descriptive terms.

Hand lay-up is descriptive of several procedures in which a single mold, either male (plug) or female (cavity) type is used. A surface skin layer called the gel coat is placed on the mold followed by the placement of various layers or reinforcements. Grooved brayers are used to compact the resin-saturated reinforcements and remove entrapped air. The lay-up normally cures at room temperature; however, heated molds or surface heaters may be used to complete cure.

Hand lay-up is popular because the process is simple, molds are low in cost, part strength is proportional to the glass content, and there are no size restrictions. Large boats (150 ft), swimming pools, automobile bodies, furniture, sinks, machine guards, amusement park equipment, diving boards, showers, septic tanks, and numerous aircraft and aerospace applications of fibrous-reinforced plastics (FRP) are formed by hand lay-up techniques. (See Figure 4-16.)

Molds may be made of wood, plaster, plaster/metal, selected polymer, or metal. All molds must be properly prepared to prevent the composite from sticking to the mold surface. Wood and plaster molds are commonly sealed with a coating of epoxy or polyester. It is important that all pores of any mold be sealed. A mold release is then applied to the mold. These mold-releasing agents may be in film, sheet, or paste form and are designed to form a release layer between the composite and the mold surface. Proper mold preparation is one of the most important functions in the molding cycle.

The gel coat may be applied by brush, roller, or spray methods. If the gel coat is properly applied, a protective, decorative, colored surface will result. Polyesters and epoxies are familiar gel coatings.

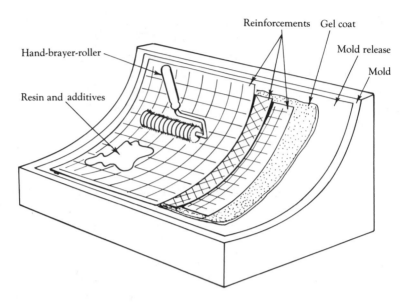

Figure 4-16. Concept of hand lay-up process.

They are normally not reinforced. This outer protective layer is normally the only surface with a smooth finish. The back side will have the reinforcement texture visible.

Performance properties may be altered greatly by varying the resin-to-glass ratio, the type of resin, the form and direction of the reinforcement, and other additives. There are several variations in the method of depositing or laying the reinforcements on the gel coat. Hand lay-up implies that little equipment is needed. In the most simple operations the resin may be mixed in a bucket and applied with a brush and squeegee. The cloth, mat, honeycomb reinforcements, or SMC may be carefully placed on the gel coat layer by hand. In all lay-up operations, it is important to bray out any bubbles entrapped between the gel coat and reinforcements. (See Figure 4-17.)

In commercial operations, hand lay-up and spray-up operations are often used alternately with the deposition of the reinforcements. Spray-up is a technique where fibrous rovings are fed through a chopper and ejected into a resin stream onto the mold. (See Spray-Up.) Fibrous glass mats are most popular in hand lay-up operations because they are easily handled, made to conform to mold shapes, and are relatively low cost. It is common to place a veil or surface

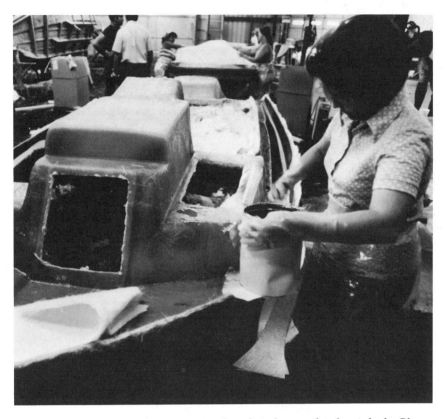

Figure 4-17. This worker is using a brush to lay-up this boat deck. Gloves are used by the worker to protect her hands. (CertainTeed Corporation)

mat next to the gel layer. This fine fiber mat will result in a resin-rich layer and a superior surface finish. The coarser mat or cloth materials are then placed over this layer. It may be desirable to alternate layers of mat, cloth, or roving materials to produce directional strength or additional strength in selected areas of the part.

This labor intensive operation is dependent on the skill of the operator in producing uniform, high-quality parts. The process is also messy and the production rate is limited (sometimes hours).

After curing and sufficient hardness have been attained, the part is removed from the mold and further trimming or finishing operations may be performed. (See Fabrication Processes.)

Spray-Up Molding

Spray-up may be considered a variation of hand lay-up. The process may be accomplished by hand or partially automated. The application gun or head is used to propel chopped fiber and liquid resin against the mold surface until the desired thickness is reached. (See Figure 4-18.) The mold preparation is essentially the same as hand lay-up. The gel coat is applied by the spray gun. After gel has been attained, spray-up of the resin and chopped fibers begins. Normally about 0.065 in. of material is deposited between roll out. Corners and rib are usually given additional layers so that the finished part can withstand stress in these areas. Roll out in hand lay-up or spray-up is important because the gel coat must not be damaged. Poor roll out can induce structural weakness by leaving air bubbles, dislocation of fibers, or poor wet out.

There are five commonly used resin-dispensing systems in the spray-up process: (1) airless two-component systems with external mixing, (2) air atomization with internal mixing, (3) airless atomization with external mixing, (4) air atomization with external mixing, and (5) airless atomization with internal mixing. In any internal

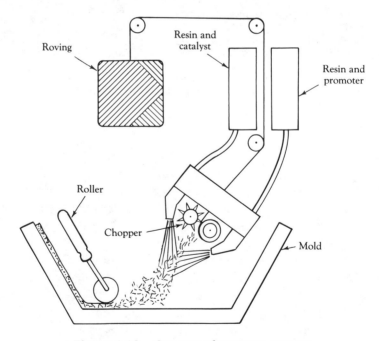

Figure 4-18. Concept of spray-up process.

mixing system, it is essential to blend only what is to be used immediately and the equipment—including pumps, hoses, and gun—must be scrupulously flushed and cleaned before cure can be accomplished.

The quality of the product relies heavily on the skill of the operator but the labor and raw materials costs are lower than hand lay-up operations. A high degree of uniformity and strength control is maintained in hand lay-up operations because preformed mats and fabrics are used. Spray-up operations may be automated and it is more practical to fabricate large parts or coat other substrates such as steel tanks. (See Figure 4-19.) Highly contoured or stressed areas

Figure 4-19. Intraply hybrid composite cylinder is being produced by combining long continuous filaments and shorter reinforcements by spray-up methods. (CertainTeed Corporation)

may be given additional thickness. Metallic tapping plates, stiffeners, or other reinforcing components may be placed in desired areas and oversprayed. Hand lay-up and spray-up operations are applied to core materials as skin layers in various sandwich composite products.

Spray-up may be done on a thermoformed thermoplastic shape. The procedure is sometimes called **rigidized-shell spray-up**. In this process, a thermoplastic sheet is thermoformed or fabricated into the desired shape. The thermoplastic material becomes the external or exposed layer replacing the gel coat. PVC, PMMA, ABS, and PC are used as skin or shell materials. Hand lay-up or spray-up operations reinforce the back side of these shapes to produce a strong composite with a thermoplastic skin. (See Figure 4-20.) Lightweight

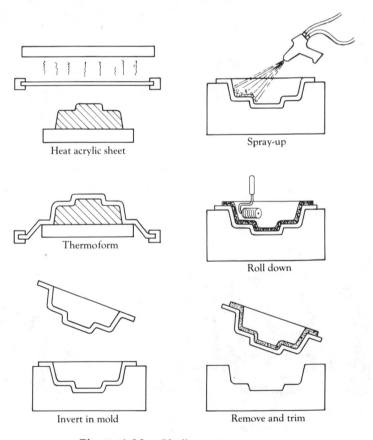

Heat acrylic sheet

Spray-up

Thermoform

Roll down

Invert in mold

Remove and trim

Figure 4-20. Shell coat process.

bathtubs, sinks, bathtub–shower combinations, small boats, van conversions, exterior signs, and car top carriers may be manufactured using this technique.

Vacuum-Bag Molding

Vacuum-bag molding is a process where a bag is placed over the lay-up (male or female mold) and a vacuum is drawn between the bag and the mold. Atmospheric pressure forces the material uniformly against the mold. Parts made by this process are dense because most voids and excess resin are removed. Large parts and superior inside (surface next to the bag) surface finishes are possible. There is excellent adhesion to composite or sandwich core surfaces.

Honeycomb materials as well as mats, fabrics, and preimpregnated forms of reinforcements are used. Compound curves (boat hull) may require a satin-weave cloth or chopped rovings. If a smooth surface is required on the exterior of a boat hull, a female mold would be selected. A sink would be formed on a male mold because the inner surface is to be smooth.

Popular wet lay-up resins are epoxies and polyesters. Prepregs may include polysulfone, polyimide, phenolics, diallyl phthalate, silicones, and other resin systems. SMC and TMC may be used. Most of the bag processes use thermal curing to attain faster cycle times and improved product properties. Infrared induction, dielectric, microwave, xenon flash, ultraviolet, electron beam, and gamma radiation may be used in the curing of some resin systems.

Because heat is required in many operations, ceramic, aluminum, or other metal tooling is used.

Mold-releasing agents may include plastic films, waxes, silicone resins, PEP, PTFE, PVA, polyester (Mylar), and polyamide films. Once the surface of the mold is prepared, a bag seal of zinc chromate paste or tape is placed around the periphery of the mold. A sacrificial ply (usually a fine, resin-impregnated fabric) is then placed on the mold surface with a flexible dam. A finely woven Dacron (polyester) or polyamide fabric peel ply is carefully positioned and the composite plies are laid in the specified designs on this surface. The edge bleeder material (burlap or other fabrics) is positioned around the mold vacuum venting system and a perforated release film is placed over the lay-up. Perforation will allow air and excess resin to escape. Release fabric of Dacron or Teflon is then used. Bleeder plies of cloth or mat are laid on the release fabric followed by a perforated breather ply of polyamide or Tedlar (PVF). These last two plies are used to prevent the bag from sealing itself against the lay-up, trapping air, or bridging sections of the mold. On some composites a caul plate is used to ensure a smooth surface and minimize variations in temperature during the curing process. The bag (silicone rubber blanket, Neoprene, natural rubber, PE, PVA, cellophane, PVC, or PA film) is positioned and the contents are evacuated under a vacuum of 25 in. Hg or about 12 psi of external pressure. To prevent the excess liquid resin from being drawn into the vacuum lines, a resin trap is used. The concept of wet lay-up vacuum-bag molding is illustrated in Figure 4-21. It is now ready to be cured in a heated chamber or by other methods. In large area applications, such as boat decks and hulls, plywood, balsa, or foams may be part of a vacuum-bag, sandwich-constructed composite.

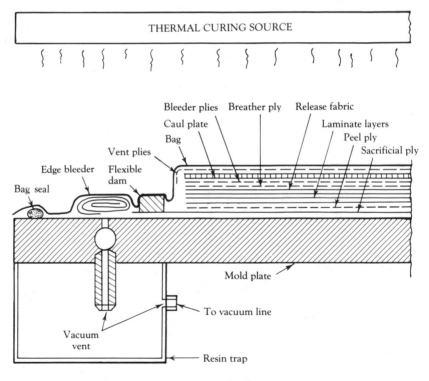

Figure 4-21. Concept of wet lay-up vacuum-bag molding.

Prepreg lay-up techniques are similar. Dry preimpregnated materials are usually more difficult to form in curved sections. Heat guns or lamps may be used to assist in laying the materials into these sections. Rollers or a squeegee may be used to ensure good contact with the mold and bag. The concept of preimpregnated lay-up vacuum-bag molding is illustrated in Figure 4-22.

Pressure-bag, rubber-plunger, rubber-bag, autoclave, and hydroclave molding are techniques used when additional density or pressure are required to form the material.

Pressure-bag techniques use air to inflate the bag and force the plies against the mold. In autoclave techniques, heated gas or steam is used. The term hydroclave implies that a heated fluid is used. In all pressure designs, the mold must be able to withstand the molding pressures. The rubber-plunger molding is a process using a male elastomer plunger to force the lay-up against the mold. It is similar to matched-die molding. Long hollow pipes, tubes, tanks, or other objects with parallel walls may use an inflatable rubber bag to force

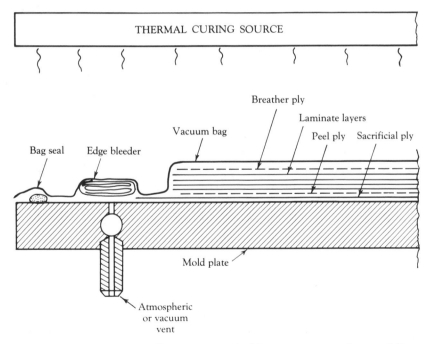

Figure 4-22. Concept of preimpregnated lay-up, vacuum-bag molding.

the lay-up against the mold surfaces. At least one end of the object must be open to insert and remove the rubber bag.

Filament Winding

Filament winding is a reinforcing process and a method of continuous fiber placement. Most fiber placement techniques may be considered preforms. This wound preform is then molded by other techniques including compression molding, pressure-bag, inflatable rubber-bag, or other processes.

Filament winding is a reinforcing process where continuous filaments are wound on a mold called the mandrel. The reinforcements are oriented in the axis of the load expected on the finished product. The matrix transfers the loads to the continuous reinforcements. Long continuous filaments are able to carry more load than random, short filaments. Filament winding tension varies from 0.25 to 1 lb per end (a group of filaments) and is critical in controlling or limiting voids and overlapping of bands.

Filament-wound applications include rocket engine cases, pressure vessels, underwater buoys, radomes, nose cones, storage tanks, pipes, automotive leaf springs, helicopter blades, spacecraft spars, fuselage, and other aerospace parts. (See Figure 4-23.)

There are two basic methods of filament winding:

1. In wet winding, the reinforcement and matrix resin are applied during the winding stage. As resin is squeezed between strands of filament, entrapped air is forced out. (See Figure 4-24.) Excess resin content, wetting, and throwing from centrifugal forces are sometimes a problem. High production ratios are of little value if the product is not of acceptable quality. The goal is to wind as fast as is practical.
2. Dry winding methods use preimpregnated B-stage reinforcements. The use of prepregs helps to ensure consistency in resin-to-reinforcement content design.

Manufacturers are the first choice in seeking reliable information on curing agents and ratios to be used in the matrix. Cure rates may vary greatly in wet or prepreg windings, and curing may be accelerated by using heat mandrels, by heating in ambient ovens, or by using hardeners that may cure from other energy sources such as ultraviolet light, high-intensity xenon lamp, or beam radiation.

(A)

BLADE WINDING

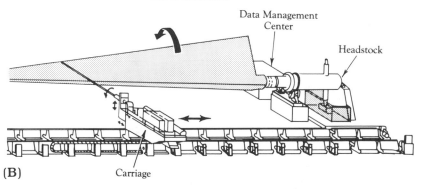

(B)

Figure 4-23. (A) Photograph of 127 ft, 13 ton filament-wound wind turbine blade. (B) Schematic of computer-controlled winding of blade. (Engineering Technology, Inc.)

Reinforcements are available in a number of forms: rovings, fabrics, tapes, mats, honeycombs, metal strips, foils, and other materials depending on design performance parameters. Such operations may require a combination of techniques (e.g., preimpregnated rovings overwrapped with fabrics and a layer of chopped glass applied to serve as a barrier against outside attack or damage).

Figure 4-24. Wet filament winding. (Owens–Corning Fiberglas Corporation)

Potentially, any continuous reinforcement could be filament wound. Over 80% of all filament winding is accomplished with E-glass roving. A major problem with glass reinforcement is its low modulus of elasticity when compared with other materials. To overcome this problem, stiffeners may be incorporated into the design. Higher modulus fibers of carbon, aramid, or Kevlar may be selected. Because these fibers are more costly, the composite may contain graphite fibers for stiffness overwrapped with less costly glass fibers. For some applications, boron fibers, high-tensile wire, beryllium, polyamides, polyesters, asbestos, and metallic ribbons have been wound. Selected properties of filaments are shown in Table 4-6. Although the monofilament strengths of E-glass and S-glass are shown to be 3.45 and 4.59 GPa, the standard deviation of $\pm 10\%$ should be considered. The theoretical strength could exceed 10–13 GPa but because of stress, corrosion from humidity, internal flaws, defects caused in processing, drawing, handling, packaging, and winding, these strengths are not attained. With better manufacturing and processing technology, the theoretical limits may be approached.

There are a number of winding methods and patterns:

1. **Classical helical winding** is something like the wrap in a skein of yarn. A helical pattern (up to 85° to the axis) is placed upon the mandrel.

Table 4-6. Selected Filament Properties

Filament	Density (kg/m³)	Tensile Strength (GPa)	Modulus (GPa)
E-glass			
Monofilament	2554	3.45	73
Strand	2547	3.34	72
S-glass			
Monofilament	2500	4.59	97
Strand	2491	3.79	85
Graphite	1772	2.24	345
Aramid	1728	2.76	131
Carbon	1760	2.76	200

2. **Circumferential winding** is like winding thread on a spool. They are sometimes call hoop winders.

3. **Polar** (also called planar) **winding** wraps from pole to pole as the mandrel arm rotates about the longitudinal axis. They are generally used to build convex pressure vessels. Filaments may be butted or lapped in various configurations.

4. The mandrel or in some instances the substrate to be reinforced, such as a thin-walled aluminum or plastic tube, may be **continuous helical wound**. The rotating frame holds the filament wraps fed by creels. The process may be continuous in the manufacture of smooth-bore pipes.

5. In **continuous normal-axial winding** the composite will have both axial and circumferential reinforcements.

6. **Continuous rotating mandrel** describes a process of placing fibers and overwrapping as the mandrel rotates and travels through the curing oven.

7. **Braid-wrap winders** use a maypole action to produce a composite reinforcement. Plastic, metal, or elastomer substrates may serve as the mandrel in a continuous reinforcing process.

8. **Loop-wrap windings** are used in the manufacture of tension straps and automobile springs. (See Figure 4-25.)

Epoxies are the major matrix system for filament winding because they have superior mechanical properties, fatigue behavior, heat resistance, lower shrinkage, and superior bonding strength. Epoxies may easily be prepregged or applied in wet operations. Polyesters are selected because they are about half the cost of epoxies and can be

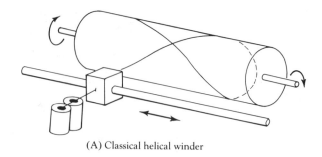

(A) Classical helical winder

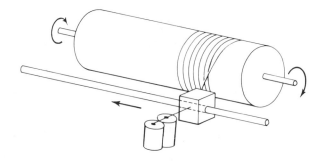

(B) Circumferential winder

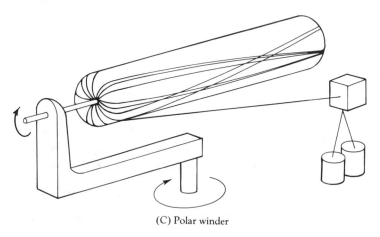

(C) Polar winder

Figure 4-25. (A)–(H) Concepts of winding methods and patterns.

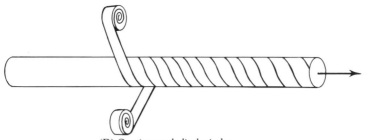

(D) Continuous helical winder

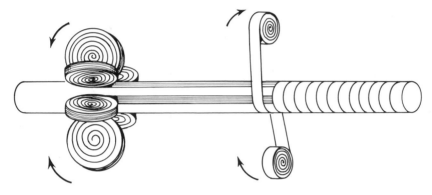

(E) Continuous normal-axial winder

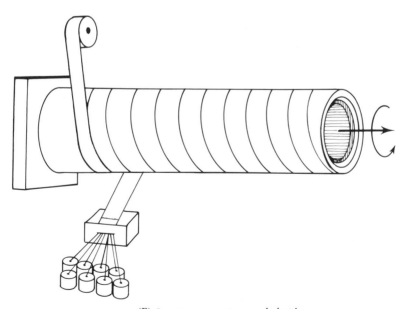

(F) Continuous rotating mandrel with wrap

Figure 4-25 Continued

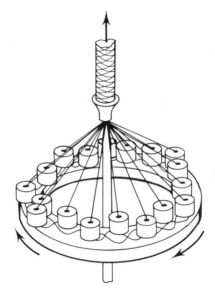

(G) Braid-wrap winder

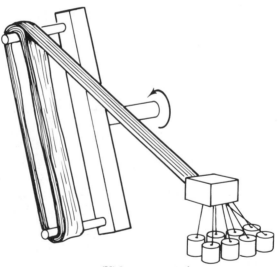

(H) Loop-wrap winder

Figure 4-25 Continued

applied in wet or prepreg processes. There are a number of promoters and initiators that advance polymerization. Silicones, phenolics, polyimides, polyamides, polysulfones, bisphenol A polyesters, and other polymers are finding increased use in filament-wound composites. Preimpregnated resin systems (thermoplastic and thermoset) offer superior quality control and eliminate the wet-out mess.

There are numerous mandrel designs. Some are designed with open ends while others permit winding of closed-end structures. They are made from a number of materials and become part of the finished product. Overwrapping of blow molded or thin-walled metal pressure tanks are familiar examples. The following mandrel construction concepts are most popular:

1. Segmented collapsible metal mandrels are relatively costly.
2. Low-melting-point alloys may be used for low-tension designs. The melt out may be reused. Some reinforced wax shapes have been used.
3. Soluble salts, plasters, and PVS—sand may be used in mandrel designs of up to nearly 3 ft in diameter. All are designed on an internal support frame and are easily washed out.
4. Some plastics and foamed mandrels may be washed out with acetone or other solvents after cure is completed. They have low strength and generally cannot withstand elevated temperature cures.
5. Inflatable mandrels are generally supported with an internal structure and usually consist of a skin layer or blow molded mandrel that is inflated and overwrapped. Only small unsupported inflatable mandrels are used because the unsupported shape will sag from the weight of the winding.
6. Slip cast plaster and plaster-chain mandrels are popular **mechanical break-out** mandrels. Pipe joints and complicated configurations use this mandrel concept.
7. There are a number of miscellaneous **variations** from cardboard or wooden tubes to hinged rectangular walls for the winding of small houses (10 × 20 × 36 ft). There are mandrels for winding aircraft wings, airframes, flywheels, and continuous pipes. (See Tooling.)

All filament windings must be removed from the mandrel (except those designed to remain on the mandrel shape). Release agents used on mandrels are similar to those used for spray-up techniques; for example, wax, PVA, and release films are common. Release tapes of Mylar (polyester), PVF, PVA, CA, PA, SI, and PE are generally half lapped to ensure a good seal. The release agent must not adhere to the mandrel or matrix if removal is to be accomplished. They must be carefully selected to prevent reaction or interference with cure of the matrix.

Continuous Reinforcing

The need for increased manufacturing efficiency and additional demand for composites have resulted in several continuous processing techniques. Nearly 10% of the total composite market is comprised of products made by continuous reinforcing. They have been described as modifications of extrusion, laminating, and lay-up molding.

Continuous processing of flat and corrugated sheet composites are familiar products used as skylights, awnings, garage doors, greenhouse panels, and wall liners or room separators. These products are made in a variety of configurations, types, sizes, and colors. Fire resistance, light transmission, and weather resistance are important considerations for most architectural applications. Continuous reinforcing of these products is achieved in a process similar to the production of SMC. A continuous stainless steel conveyor system is usually used. Carefully selected resin systems and other additives are placed on a bottom carrier film. The first carrier film creates the surface finish of the sheet. Carrier films may have a smooth, patterned, or matt finish. In some lines, a gel coat or decorative layers may be placed directly on the carrier film. Reinforcement materials are fully wetted and squeezed between rollers and a top release film. This mechanical roller action helps to compact the composite, assists in wet out, forces out air bubbles, and determines the product thickness. A series of forming dies or shoes are used to create the desired configuration as the composite is pulled through the curing area. Upon cure, the carrier films are rewound and may be reused. Continuous reinforcing is illustrated in Figure 4-26. In some products the carrier film (Cellophane) is not removed and becomes a protective layer for shipment and handling. Sheets have been made as wide as 9 ft and as long as practical. Thin composites are sometimes rolled in 100–1000 ft rolls.

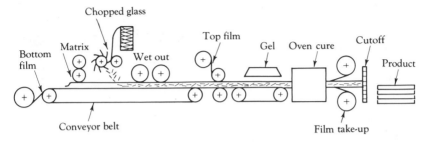

Figure 4-26. Continuous reinforcing of sheet product.

Pultrusion is a term used to distinguish the process of pulling (as opposed to pushing) a composite material through a die orifice. Although the process is associated with pulling resin-impregnated rovings through a forming die, thermoplastic materials may also be used. This latter process uses both the hot melt extrusion and pulling action of the composite.

Continuous pultrusion (thermoplastic and thermoset) and pulforming are used in the manufacture of automotive springs, chemical gutters, frames, airfoils, hammer handles, skis, tent poles, golf shafts, fishing poles, ladders, tennis racquets, vaulting poles, aerial booms, and numerous profile shapes such as I-beams, rods, and tubes. (See Figure 4-27.)

Polyesters and epoxies are commonly used matrices in pultrusion processes. Emulsion polymerization techniques have been employed using few thermoplastics. The production lines are normally horizontal but vertical arrangements are used.

The process begins by pulling continuous reinforcements through the resin bath. Glass rovings, mats, cloths or carbon, graphite, aramid, and other continuous reinforcing materials are used. Some are used in combinations and may contain SMC or wound preforms to improve omnidirectional properties. Parallel orientation of reinforcements results in a very strong composite in the direction of the fibers. Wet-out aids such as rollers or pressure impregnators are used in the impregnating bath to ensure full wetting of the reinforcement. Excess resin is removed to expel any air and to compact the composite. The resin-impregnated material is then passed through preforming dies (mechanical aids) to ensure that the reinforcements are carefully aligned. The product then passes into the curing section, where chrome plated steel dies may be used for thermal and

(A)

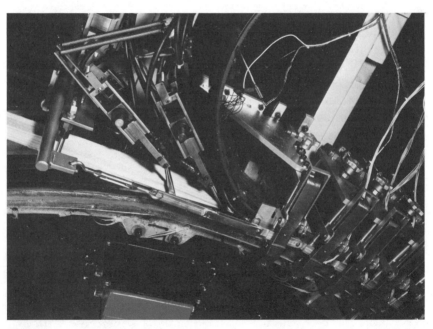

(B)

Figure 4-27. Pulforming of a curved composite leaf spring: **(A)** glass rovings feeding through wet-out tank; **(B)** heated belt die closure cure section; **(C)** spring stock exiting the die/belt section; and **(D)** flying spring stock cut-off saw. (Goldsworthy Engineering)

(C)

(D)

Figure 4-27 Continued

chemical curing. If radio (70 MHz) or microwave (2500 MHz) frequencies are used, a PTFE nonconducting die is needed. The abrasive action of reinforcements and matrices wear die surfaces quickly. For flat plate stock, a process similar to glass production has been used. The pultrusion is pulled over molten tin, with the tin bed serving as the heat source and the die surface.

The continuous cure must be carefully controlled to prevent cracking, delamination, incomplete cure, or sticking to the die surfaces.

Simple belt or caterpillar tread pullers provide the pulling forces needed in this system. A typical pultrusion process is shown in Figure 4-28.

Pulforming is a modification of the pultrusion process. The process uses a number of forming devices (molds) to shape the pultruded (pulformed) composite into curved or bent products or into products with varying cross-sectional shapes. This is accomplished in several ways. In one method, rotary or circulating male and female dies are brought together on the pultruded form and the composite is cured. Hammer handles are a familiar example. The process may be intermittent or continuous in the forming operation. In another technique, female or male dies may be used to bend or shape the curing composite. Large circular composites may be formed by placing the pultruded material in a large rotating mold with a female cavity. A flexible metal band is placed over the material forcing it into the cavity for final cure.

Pultrusion of thermoplastic composite materials may be accomplished by modification of a manifold die on conventional extruders.

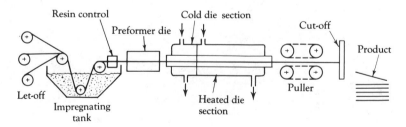

Figure 4-28. Pultrusion of resin-impregnated reinforcing materials.

Cold Molding

This process should not be confused with cold forming. Both terms are used to indicate that the material is formed in an unheated mold or die. Cold press molding is more descriptive of the technique where reinforced compounds (organic or inorganic) are pressed in unheated matched dies. The pressed shape is then cured in an oven. Steam atmosphere may be used in the ovens to harden or cure formulations containing cement or plasters. Typical applications of these heavily filled and reinforced composites are wall and floor tile, electrical insulator parts, utensil handles, and bobbins.

Cold Forming/Stamping

This process is sometimes called cold stamping. The process involves heating reinforced thermoplastic sheets and then forming between matched dies until cool. Like cold molding, cold forming may be accomplished in compression of stamping presses.

The reinforced thermoplastic sheet may be made by continuous pultrusion or by alternately compressing layers of reinforcements, films, adhesives, and emulsions into a sheet form. Even honeycomb sandwich construction using latent curing systems has used the cold forming process. The surface may be embossed or decorated with overlays of film, metal, fabrics, foams, or other combinations.

Most of the applications have been concealed. Fan guards, seat backs, trim pads, fender liners, tank covers, and exhaust and air treatment ducts in automobiles are familiar examples. These parts are light, strong, and corrosion resistant and may be formed rapidly on inexpensive dies. Parts may be molded with ribs and varying cross-sectional thickness.

Sintering

Sintering of polymers has been adapted from sintering operations used by the powder metallurgy industry. Composite-, cellular-, and alloy-type materials may be made by this process. Most are made of powders pressed at a temperature just below the melting point of the polymer. Particles are fused (sintered) together, but the mass, as a whole, does not melt. Polyimides, PTFE, PA, and other specially filled materials are used. After sintering, some parts are postformed by further heat and pressure application. Some cellular materials have been produced by placing glass spheres in the powder formulation before sintering, while others have been produced in a similar process by placing various salt (sodium sulfate) crystals in the mix. After sintering, a solvent solution is used to dissolve the crystals, leaving a porous polymer. The process is sometimes called leaching. It is possible to alternate layers of compatible polymers, reinforcements, and soluble crystal mixtures to produce a true composite component.

Most applications are for bushings, bearings, hubs, and electrical insulation. Sintering is the principal method by which PTFE is processed.

Liquid-Resin Molding

Liquid-resin molding (LRM) is sometimes called liquid transfer molding, liquid injection molding, or resin transfer molding (RTM). It should not be confused with RIM. LRM fills a gap between hand lay-up, spray-up, and SMC matched-die molded composite parts. The process uses liquid resins (reactive polyesters and epoxies). Preforms of fibers or other reinforcements are placed into the mold cavity prior to forcing the liquid resin into the mold. The resin is forced into the mold cavity and cured. The low pressures do not distort or move the desired fiber orientation. LRM is an important embedding process used to pot delicate wires and electrical components into one integral component. LRM is usually associated with the production of large parts, such as boat hulls. Boat hatches, computer housings, and fan shrouds on tractors may be molded by this process. The basic concept of LRM is illustrated in Figure 4-29.

Vacuum-Injection Molding

This process is a variation of LRM. Preforms are placed on a male mold. After the female mold is closed, a vacuum is drawn, pulling the reactive resin and additives into the mold. The concept is shown in Figure 4-30.

Thermal Expansion Resin Transfer Molding

Thermal expansion resin transfer molding (TERTM) is a variation of the RTM process. Preformed cellular (PVC or PU) cores (like a mandrel) are wound or wrapped with reinforcements. The sur-

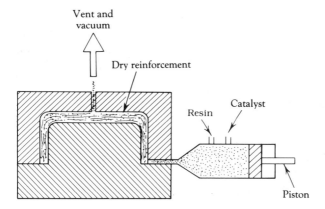

Figure 4-29. Concept of LRM.

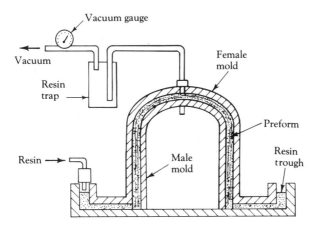

Figure 4-30. Vacuum-injection molding concept.

rounded cellular preform is then placed into a matched die. Epoxy resins are injected to impregnate the reinforcement. Heat causes the foam to expand further, generating pressure against the impregnated reinforcements and expelling any excess matrix.

Casting Processes

A number of composite structures and components may be produced by casting. Casting processes are similar to those used in the glass, ceramics, and metals industries. The essential difference between casting and molding is that only atmospheric or centrifugal forces are used to shape the product. Most molds must be carefully prepared and coated with mold-releasing agents to ensure easy removal. The matrix may consist of hot melts, monomers, solvent solutions, powders, or other modified monomer resins. Casting has been a favorite process for many years because it is generally the simplest and presents a low-cost, short-run potential not possible with most molding techniques. Parts have little or no stress or orientation. Rotational casting produces one-piece hollow objects!

Casting processes may include the following six distinct groups: (1) simple, (2) film, (3) hot-melt, (4) slush, (5) rotational, and (6) dip.

Simple Casting

Simple casting consists of pouring liquid resins or molten plastics into molds and allowing the mixture to polymerize or cool. The process generally uses open molds similar in design to those used in making candles or baking cakes. Heavily filled or reinforced polymers may be cast into blocks, tubes, or other shapes and machined to the desired shape. Some cellular plastics are produced by casting but are included under expanding processes. (See Expanding Processes.) Composite toolings and fixtures are cast into desired shapes to aid in the processing of other materials.

Casting is an important embedding process. Embedding is a process of complete encasement of an object in a polymer matrix. It is achieved by several techniques. **Casting** is accomplished by pouring the resin mix into a mold. The liquid completely surrounds the object. After cure is completed, the encapsulated casting is removed from the mold. Clear decorative cast embedments containing coins, insects, or rocks are familiar examples. **Potting** is a similar process but the mold becomes part of the product. Electrical wires are potted inside a metal shell in some transformer designs. **Impregnating** is used to soak or wet out the assembly (embedment) completely. The impregnated assembly is then cured. Vacuum or centrifugal forces are sometimes used to assist in the wetting. The process may be considered a coating operation. Impregnating spools or electrical wires are typical applications. **Encapsulating** is a process where the object is dipped into a thick, viscous material and cured. The coating is relatively thick. Individual components or complete assemblies may be coated. After potting, many components are encapsulated.

Liquid-resin molding is used to force liquid materials into a mold containing the embedment. (See Liquid-Resin Molding.)

Simple casting molds may be made of wood, plaster, metal, glass, or selected polymers. Thermoformed PE molds are commonly used, but PUR and SI molds are preferred for casting intricately designed furniture components.

Probably the best known casting resin is polyester. Epoxies, phenolics, acrylics, silicones, polyamides, cellulose acetate butyrate, and polyurethanes are commonly cast. Acrylic monomers are cast between two polished sheets of glass. The glass sheets separated by a gasket serve as a mold. After heating and complete polymerization, the cast acrylic sheet is removed from the mold. Many profile shapes are cast. Some polymers may be produced by continuous casting between stainless steel belts, with polymerization occurring between these two moving belts. (See Continuous Molding.)

Familiar cast objects include jewelry, cast sheets for glazing, knobs, table tops, and furniture parts. Only a few composites are produced by simple casting. Most examples are heavily filled or reinforced pieces used as forming fixtures and tooling. Cast PA are used for bearings, gears, and bushings. Reinforced PA composites find uses in structural applications requiring heavy ribs and bosses.

Film Casting

Film casting is a process of dissolving polymers in a solvent or using emulsions of a polymer and placing them on a continuous stainless or polymer belt or large roller. The solvents are evaporated and the film is left as a deposit on the belt. This film is then wound on a takeup roller. CA, CAP, PVC, PVA, PMMA, and other selected polymers may be film cast. (See Laminating.)

A typical application for PVDC, PVC, PMMA, and ABS is the packaging of food and pharmaceuticals. Tentering (frame attached to two sides used to stretch polymer) is used to produce heat shrink films for packaging.

Hot-Melt Casting

Hot-melt castings are used as strippable coatings, adhesives, or embedding materials. Hot-melt formulations may be made of nearly all melt processible polymers. Ethyl cellulose, CAB, PA, and PE are common hot-melt polymers.

Slush and Static Casting

Slush and static casting involves pouring a dispersion of PVC or powders into a mold. The mold is heated until the desired wall thickness is reached. In slush casting the process usually begins with heated molds. Excess materials are drained, leaving the polymer clinging to the mold walls. The mold is returned to the heating chamber until all polymeric materials are fully fused together. After a cooling cycle the part is removed. The process is similar to slip casting of ceramic materials, except heat is used to cause the polymer to fuse to the mold sides. In a related process, vibrational microlamination (VIM), a combination of heat and vibration is used. (See Figure 4-31.) Reinforcing fibers may be used to strengthen the matrix but wicking may present some application problems. Thin layers of homopolymer alternated with reinforced layers are used to overcome this problem.

Huge storage tanks, hollow toys, syringe bulbs, or other containers may be made by this technique.

Rotational casting (sometimes called centrifugal casting) is similar to slush casting. Selected thermoplastic and thermosetting materials may be rotational cast. The process is sometimes called (incorrectly) rotational molding.

Powders, monomers, or liquid dispersions are measured and placed in the multipiece mold, which is generally made of aluminum. During the heating or curing cycle, the mold is rotated in two perpendicular axes simultaneously until the liquid or powder materials have formed a skin layer on the mold surface. Cast crystalline polymers are generally air cooled while amorphous polymers may be cooled quickly by a water spray or bath. Once cooled, the part is removed and the process is repeated.

True composites may be produced by this method. Short rein-

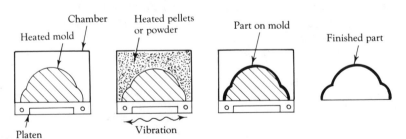

Figure 4-31. VIM process.

forcing fibers may be used to increase strength, but fiber loading seldom exceeds 20%. Foam-filled and double-walled items may be produced. Multilayered walls of compatible polymers can be used. In one process a solid outer skin layer is produced followed by the release of a second charge of material from a *dump box*. This charge may contain a different reinforcement and matrix from the skin layer.

The wall thickness may be controlled by rotating one or the other axis at a different speed. This throws more material against the mold surface in the faster axis. This concept is shown in Figure 4-32.

Tubular composites are made by rotating the mold on one axis. A hollow mold is prepared and prepregs or more commonly wet lay-ups are placed on the mold wall. Centrifugal force causes the reinforcement and matrix to compact. The mold is heated to complete the cure and the composite is removed. Typical products include pipes, ducts, submarine launcher tubes, fuel tanks, and parabolic shapes.

Inflatable mandrels may be placed in the tube shapes to densify the lay-up, provide a smooth inner surface, or place special ribs or other geometric designs on the internal surface of the product.

Dip Casting

Dip casting is a process of heating a mandrel and dipping it into a plastisol dispersion. The heated tool or mandrel remains in the dispersion until sufficient material has collected on the surface. The

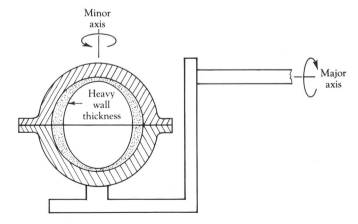

Figure 4-32. Concept of rotational casting.

mandrel is then removed from the dip tank and placed into a heating chamber until all dispersion particles have completely fused together. Upon cooling, the part is then stripped from the mandrel. Several layers or alternate colors and formulations may be applied by alternately heating and dipping. This process should not be called dip molding and is not to be confused with dip coating. (See Coating Processes.)

Squeeze coin purses, spark plug covers, some overshoes, and surgical gloves are familiar examples of products made by the dip casting process.

Thermoforming Processes

Laminar, reinforced and particulate composites are produced by thermoforming techniques. The process uses mechanical aids, matched dies, and air or vacuum pressure to shape a heated thermoplastic sheet.

Extruded, calendered, laminated, cast, and blown films or sheets are heated by convection, conduction, or radiation, with electrical radiant heating being the most common method. Cooling is accomplished by conductive losses to the air and tooling.

Aluminum molds are the most common tooling materials, but steel, wool, plaster, and selected plastics are also used. Steel dies and inserts are used for punch and in-mold trimming.

Male and female molds are used in a variety of thermoforming techniques. Vertical sidewalls may be used in female mold designs because part shrinkage allows easy part removal. Male molds require sufficient draft to ensure stress-free part removal. Matched-mold forming and plug-assisted forming use male and female molds in combination.

Most thermoplastic material may be used in continuous roll- or extruder-fed equipment. There is considerable energy, time, and space savings by directly thermoforming sheets as they leave the extruder.

Typical continuous roll-fed thermoformed products include ice cube trays, ducts, instrument panels, furniture parts, bottle base cups, tote trays, toys, food service trays, and numerous food containers, such as margarine tubs, dairy cups, and bakery item display covers.

Polystyrene is the most commonly thermoformed material. It is made into single portion package food containers for deli goods, cake covers, disposable dinnerware, and soft drink cups. Cellular PS is made into meat trays and egg cartons. ABS, PVC, PP, PE, acrylics, and other polymers are selected for barrier applications in packaging foods, cosmetics, and pharmaceuticals. Jellies, butter, and cookies are sometimes packaged in transparent blister packages. Skin packaging requires no mold. The hot film is formed over the product. Seven-layered retortable, microwavable soup bowls and trays are thermoformed. Crystallized PET (CPET) is used for dual-ovenable food trays. Liquid crystal polymers (LCP) are finding a growing market in aircraft interior compartments, machine cases, trays, and other housing components.

Large, heavy-walled products or low-volume thermoforming is generally done on sheet-fed thermoforming equipment. Both shuttle and carousel or rotary equipment are used.

Typical sheet-fed thermoformed products include canoes, boat hulls, signs, bodies for recreational vehicles, refrigerator door liners, and automotive and aircraft components.

Thermoforming techniques may include variations of straight vacuum, drape, matched mold, pressure-bubble plug-assist vacuum, plug-assist vacuum, plug-assist pressure, vacuum snapback, pressure-bubble vacuum snapback, trapped-sheet contact-heat pressure, air slip, free, and mechanical.

In a special thermoforming process, biaxially oriented parts of ABS, PC, UHMWPE, PP, HDPE, and others are formed. The blank (extruded, compression molded, sintered powders) is heated just below its crystalline melting point and stretched (pressed) into a sheet form. The (preform) sheet is then transferred to the thermoforming press and a plug stretches the hot material. Air pressure (80–100 Pa) is applied to force the hot material against the mold sides to complete the forming. The process is similar to plug-assist forming but is called solid-phase pressure forming. Biaxial orientation enhances the strength, toughness, and environmental stress-crack resistance of polymers. (See Figure 4-33.)

Straight Vacuum

Straight vacuum forming results when a heated sheet is drawn (forced) into a female mold cavity. Areas of the sheet reaching the mold surface last are thin. This concept is illustrated in Figure 4-34. Female molds are preferred for multicavity forming. Male molds

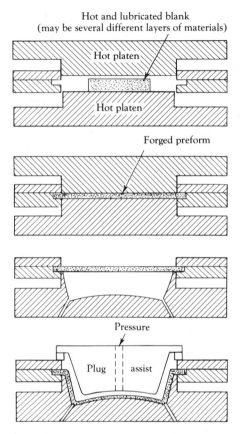

Figure 4-33. Biaxial orientation by solid-phase pressure forming.

require more spacing. Draw or draw ratio is the ratio of the maximum cavity depth to the minimum span across the mold opening. It is best if this ratio does not exceed 0.7:1 for HDPE.

Drape

Drape forming is accomplished by drawing a heated sheet over a male (positive) mold. High draw ratios are accomplished with drape forming. Ratios of 4:1 are commonly attained. Any scratch or other mold mark-off (marks from the mold) will be visible on the inside (side next to the mold) of the part. In straight vacuum formed parts, the mark-off will be on the outside. Webbing may occur if a space equal to the mold height is not left between male forms. (See Figure 4-35.)

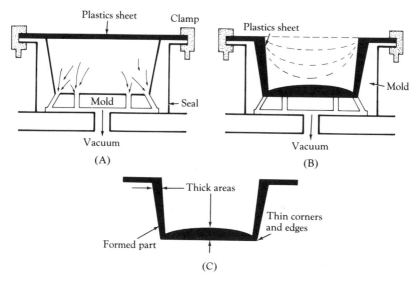

Figure 4-34. Straight vacuum forming. (Atlas Vac Machine)

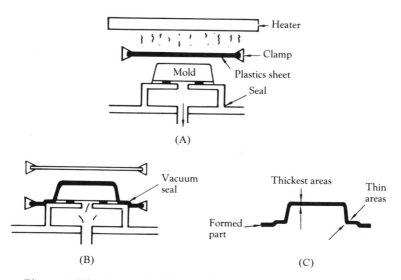

Figure 4-35. Principle of drape forming. (Atlas Vac Machine)

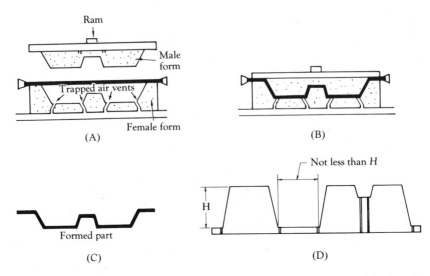

Figure 4-36. Principle of matched-mold forming. (Atlas Vac Machine)

Matched Mold

Matched-mold forming is similar to compression molding. The heated sheet is formed between male and female molds. There will be mark-off on both surfaces but excellent molded detail and dimensional accuracy can be obtained. Patterned surfaces such as pebble grain, wood grain, or simple sand blasting are recommended with PE materials to prevent air pocket mark-off. Figure 4-36 helps to illustrate matched-mold forming.

Pressure-Bubble Plug-Assist Vacuum

Pressure-bubble plug-assist vacuum forming is used for deep draw ratios in a female mold. Air is used to force the hot sheet to bubble and stretch to the predetermined height. A warm plug is then activated, forcing and further stretching the bubble into the mold cavity. Plug penetration is usually from 70 to 80% of the cavity depth. Air pressure from the male mold and vacuum from the female mold are simultaneously applied. The remaining material is then forced against the mold side, completing the forming process. Vacuum or pressure alone may be used depending on product design. This forming process is shown in Figure 4-37.

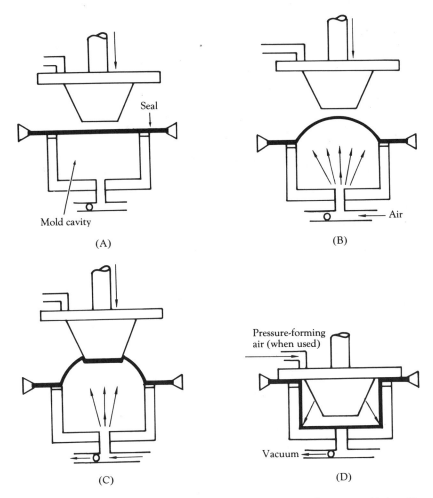

Figure 4-37. Pressure-bubble plug-assist vacuum forming. (Atlas Vac Machine)

Plug-Assist Vacuum

Plug-assist vacuum or pressure forming uses a plug to stretch the hot sheet mechanically and to minimize corner or periphery thinning of the product. The plug should be from 10 to 20% smaller in length and width than the cavity. It is the plug size and shape that are primarily responsible for the ultimate material distribution. Vacuum and/or air pressure is then used to complete the cycle, forc-

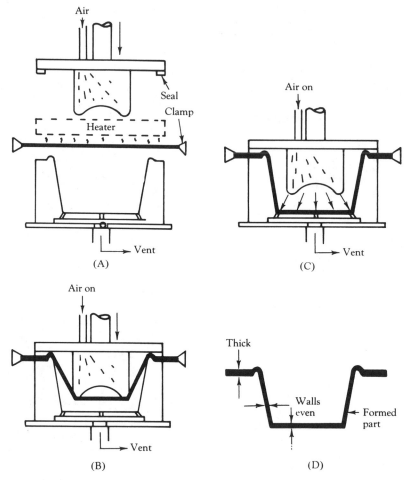

Figure 4-38. Plug-assist pressure forming. (Atlas Vac Machine)

ing the soft sheet against the mold surface. This thermoforming technique is shown in Figure 4-38.

Vacuum Snapback

Vacuum snapback forming begins by causing the hot sheet to be drawn and stretched into a vacuum box. A male mold is then lowered and the vacuum is released from the box. This causes the heated material to snap back around the male mold. A vacuum may also be drawn on the male mold to assist in the process. The process is shown in Figure 4-39.

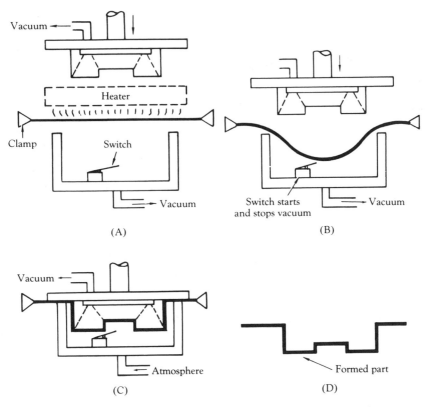

Figure 4-39. Vacuum snapback forming. (Atlas Vac Machine)

Pressure-Bubble Vacuum Snapback

Pressure-bubble vacuum snapback forming begins by blowing a bubble to stretch the material about 40%. The male mold is then lowered and vacuum is applied to the male mold. Air pressure is forced into the female cavity. This combination causes the hot sheet to snap back around the male mold. This process allows deep drawing and the formation of complex parts. Figure 4-40 shows the steps of this process.

Trapped-Sheet Contact-Heat Pressure

Trapped-sheet contact-heat pressure forming is similar to straight vacuum forming except both vacuum and pressure are used as illustrated in Figure 4-41.

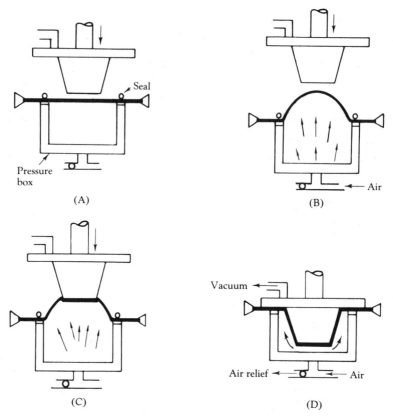

Figure 4-40. Pressure-bubble vacuum snapback forming. (Atlas Vac Machine)

Air Slip

Air-slip forming is a process similar to snapback techniques except in the method of forming the stretch bubble. This concept is illustrated in Figure 4-42.

Free

Free forming is used to form a bubble up through a silhouette of a female mold. Air pressure forces the hot sheet to form a smooth bubble. The shape is determined by the height of the bubble and the silhouette mold shape. Skylight panels are familiar product examples of this process.

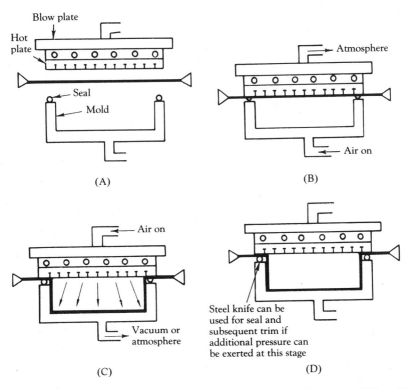

Figure 4-41. Trapped-sheet contact-heat pressure forming. (Atlas Vac Machine)

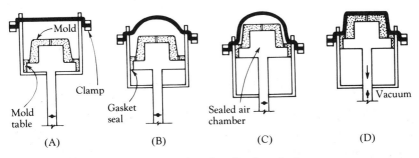

Figure 4-42. Air-slip forming.

Mechanical

Mechanical forming is sometimes classed as a postforming operation. Numerous shapes may be formed with simple forming tools or jigs. Vacuum and air pressure are not used. Only the mechanical force of bending, stretching, or holding the hot sheet is used.

Plug and ring forming is sometimes classed as a separate forming process. This mechanical forming process will result in mark-off anywhere the female or male mold touches the hot polymer.

Expansion Processes

There are a number of expansion processes and techniques to produce a polymer with a mass of bubbles or cells within the polymer or matrix. The literature is full of definitions and terms. Foamed, expanded, frothed, cellular, sponge, blown, bubble, and syntactic plastics and polymers are described. In this text, the process of making low-density materials from solid or resin matrices will be referred to as an expansion process. The resultant material will be a cellular (from Latin **cellula**, meaning small cell or room) polymer. Cellular materials may be classified as closed-cell or open-cell types. If each cell within the polymer is a discrete, separate cell (bead board insulation), it is a closed-cell polymer. If the cells are interconnected with openings between cells (spongelike), the polymer is an open-cell polymer.

Nearly all polymers (thermoplastic and thermoset) can be made cellular. Most are made into rigid, semirigid, and flexible forms. The ASTM classifies cellular materials as either rigid or flexible.

The term **structural foams** is used to refer to rigid cellular materials with an integral skin.

Cell initiation, growth, or stabilization is accomplished by a wide variety of expanding processes. The concept of density reduction (making the matrix cellular) falls into one of the following three processing groups: (1) physical, (2) chemical, and (3) mechanical.

Reinforced cellular materials have been produced with particulate and fibrous reinforcements dispersed in the polymer matrix. In physical formulations, reinforced epoxies have been mixed with pre-expanded beads (PS, PVC) and fully expanded and cured in a mold. Chemical methods involve mixing the reinforcements with the resins, catalysts, and blowing agents and allowing the expansion to occur. Short fibers tend to orient parallel to cell walls, giving improved rigidity. In one process a resin-impregnated reinforcing mat and matrix expand together, forming a reinforced-cellular product. Mixing small, hollow balls in the matrix is a mechanical dispersion technique used to produce reinforced, closed-cell materials. Cellular materials are also used in the manufacture of sandwich composites. (See Sandwich and Continuous Reinforcing.)

Strengths and moduli vary with density, polymer composition, and cell size and shape. Glass-filled epoxy syntactic materials have excellent compressive strengths.

Physical Processing

Physical processing uses a blowing agent such as pentanes, hexanes, halocarbons, or mixtures of these agents in particles of plastics. PS, SAN, PP, PVC, and PE are made cellular by this process. Polyolefins are generally cross-linked by radiation or chemical means because PE and PP experience cellular collapse before they can be cooled or stabilized. As the particle, bead, or granule of plastics is heated, the polymer becomes soft and the blowing agents vaporize. This produces a lightweight, expanded material sometimes called the prepuff, preform, or pre-expanded bead. These prepuffs are then slowly allowed to cool to prevent cell collapse. Sudden cooling may create a partial internal vacuum in each cell or prepuff. Prepuffs are generally aged 8–24 h to prevent collapse or shrinkage. The prepuffs are then placed in a mold and exposed to steam. Other energy sources have been used. This second expansion causes the prepuffs to expand further and fill all spaces and fuse together. Aluminum molds are common with molding pressures of less than 30 psi (pressure from expansion). Once the prepuffs become one integral piece, the part is cooled, resulting in closed-cell products. Lubricants or fluoroplastic-coated molds are used as releasing agents. Water spray or bath systems are used to speed the cooling cycle. Cooling time consumes 50% of the molding cycle. Further expansion may occur if the part is demolded before it has had time to cool adequately. PP and PE may require postmolding oven treatment to maintain dimensional stability.

Cellular materials are also produced by mixing a blowing agent or gas directly into the polymer melt. Extruders or injection molding equipment is used in continuous operations. Special screw shaft seals are used to prevent the escape of blowing gas. As the melt

Figure 4-43. Three extruders may be used to produce this pipe with a cellular inner core and a different exterior and interior skin. (Atochem)

leaves the die orifice or enters the mold cavity, the blowing agents vaporize, causing the polymer to expand. Blowing agents may be in powder, granule, liquid, or gaseous form. They remain compressed in the melt until they are forced into the atmosphere, where decompression (expansion) occurs. PS, CA, PE, PP, ABS, and PVC are commonly expanded by this process. Stabilization after molding is accomplished by cooling the matrix below the glass transition temperature.

Integral skinned cellular polymers are molded by extrusion and injection techniques. (See Figure 4-43.)

Continuous extrusion cellular materials are laminated with paper, fabric, foil, and other materials. (See Laminating, Coating.) Coextruded multilayered pipe, sheets, or other profile shapes are produced. These products may have a solid skin outer layer and a cellular center. They are used for flotation devices, packaging applications, and insulated pipes.

Blowing agents must be carefully metered to prevent boiling prior to emergence from the die and to control density reduction. Cooling and forming dies are used next to the die to control stabilization, produce a smoother surface, and control expansion. Most extruded cellular sheets are aged (24 h) before fabrication into other products. Continuous production, thermoformed containers for bottle covers, produce trays, egg cartons, and snack food trays are the exceptions. Dimensional stability attained by allowing cells to reach equilibrium pressure is not critical.

Profile shapes may be produced with a solid skin with progressively lower density centers. Hot melt is forced around a fixed torpedo. The shape is determined by the sizing die. As the melt expands, the skin is formed by the sizing and cooling dies. The space left by the torpedo is also filled. This concept is shown in Figure 4-44.

A large market for extruded cellular products comes from the textile and flooring industries. Molten plastics containing the expanding agents are extruded through the die orifice into continuous sheets or other shapes. These products are open celled and have no surface skin.

Several injection molding techniques are used to produce parts with a cellular center surrounded by an integral solid skin.

In **low-pressure** (100–800 psi) **injection molding**, the mold is partially filled with the melt. As the blowing agents expand the matrix and fill the mold, a skin is formed as the gas cells collapse against the mold sides. Skin thickness is controlled by the amount of melt injection into the mold, the blowing agent, and mold temperature. Low-pressure techniques produce closed-cell parts that are nearly stress free. Parts have a characteristic surface swirl pattern produced by the collapse of the cells on the mold surface.

In one **high-pressure technique**, the mold cavity is completely filled with the melt. Expansion begins when the mold is opened slightly to allow for controlled expansion. In another technique, called gas counterpressure, gas is forced into a sealed mold cavity. The melt is then forced into the cavity against the gas pressure. Gas pressure in the cavity prevents premature expansion. Once the mold is vented, expansion occurs. Both high-pressure techniques produce parts that have less prominent weld lines and that are generally free of swirl patterns.

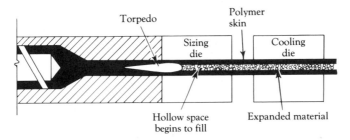

Figure 4-44. Extrusion of cellular profile shape with solid skin.

Multiple component, **coinjection molding** techniques are used to produce sandwichlike cellular materials with solid skins. Rigid structural parts or flexible skins and cellular cores are made into unique composites with hybrid properties. (See Coinjection Molding.)

A process using gas channels in the hot polymer melt is basic for a unique expansion technique. Gas is carefully injected in channels with the melt as it enters the mold. As the gas expands in less viscous, thicker sections of the melt, the polymer is forced against the surface of the mold. There is no laminar flow or swirl marks. Hollow parts (gas channels), which extend to the extremities of the piece, act as box sections, aiding the rigidity and strength-to-weight ratio. This process may be used to produce car parts, computer enclosures, and domestic goods.

Reaction injection molding (RIM) and reinforced reaction injection molding (RRIM) have an important role in the development of innovative expanding technologies. Since 1970, the process has grown rapidly with extensive use in automotive and other structural applications. The process is a modification of injection molding. (See Reaction Injection Molding.) Extruders are used to force two or more materials into an accumulator (mixing chamber). A plunger then forces the materials into the mold cavity. The reaction mixture then expands to fill the mold cavity. Parts are produced with an integral, solid skin and a cellular core. The density of the part is determined by the amount (20–65% mold volume) of material metered into the accumulator. Because clamping and molding pressure are low (50–100 psi), relative low-cost tooling may be used. Heated molds help control the skin quality and thickness. Heat also helps to reduce the molding cycle time for some polymers' formulations to less than 2 min.

Greater demands are being made for structural RIM parts. RRIM and mat molding RIM (MM/RIM) have been used to improve dimensional stability and modulus. (See Figure 4-45.)

Long fibers in MM/RIM greatly improve impact strength. The process is similar to resin transfer molding. Reinforcing mats are placed in the mold and a mixture of reactive components is forced into the cavity.

Chemical expanding processes are versatile methods of producing cellular materials as well as expandable materials from condensation polymers. In chemical expanding, liquid mixtures of resin,

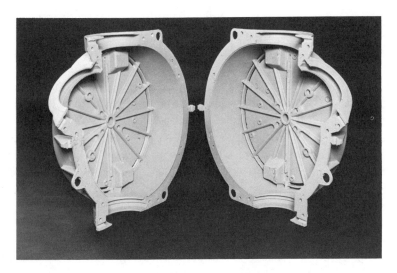

Figure 4-45. This RRIM pneumatic air conveyor valve contains 15% by weight glass reinforcement. (Hercules Inc.)

blowing agents, and other additives are joined and poured or forced into a mold. A chemical reaction occurs within the mixture and the matrix expands, curing into a celled structure. Most elastomers can be made into either open- or closed-cell materials. As the polymer expands, cell walls become thin and rupture, resulting in a matrix with a catacomb structure.

Polyurethanes, polyethers, urea–formaldehyde, polyvinyls, phenolics, polyesters, silicone, isocyanurate, carbodiimide, acrylics, epoxies, and most elastomers have been made cellular by chemical means. Some are simply cast (or expanded-in-place) into open or closed molds. Materials may be combined in accumulator or mixing heads and forced into a mold in a process similar to RIM. Resins and reactants may be mixed (internal or external) by the aerosol or pneumatic action of a spray head. Various composite systems and formulations are available. Reinforcements have been mixed in the matrix before molding or placed in molds before dispensing the resin and reactants into the mold.

Tooling for chemical expanding may use low-cost silicone- or epoxy-filled tooling. Long-run, high-quality products are generally produced in aluminum molds.

Applications for chemically expanded materials include flotation devices, sponges, mattresses, safety cushioning, insulation, and core stock for sandwich or laminated panels.

Polyurethane has been the most important material made by chemical expansion. Isocyanate and resin mixtures of polyol or a polyol blend, catalysts, surfactants, flame retardants, and blowing agents are used. In some systems water reacts chemically with isocyanate to liberate carbon dioxide. Some hydrocarbon blowing agents are activated by the heat of chemical reaction. Flexible polyurethanes are open-cell materials used extensively for comfort cushioning. Rigid polyurethanes have closed-cell structures. Polyurethanes and polystyrene account for more than 90% of all insulation for refrigerators, freezers, cryogenic tanks, and roof insulations.

Isocyanate is a popular insulation material made by carefully metering streams of isocyanate, resin, and catalyst between layers of aluminum, felt, steel, wood, or gypsum facings. The low thermal conductivity of low-density cellular polymers is a major reason for their selection as insulating materials!

Expand-in-place and spray-up techniques have also been used in the production of phenolic cellular products. Halocarbons are usually blended into the resol to produce an open-cell structure. Continuous production of slab stock is similar to that for polyurethane and isocyanurate. Thermal insulation for sandwich and panel applications is most popular.

Simple casting, rotational casting, spray-up, and matched-die molding are used in the production of cellular polyester.

Selected properties of expanded polymers are shown in Table 4-7.

There are several **mechanical** means to produce cellular polymers. For example, mechanical agitation is used to whip air or other gases into the resin or melt, while the catalysts in the resin stabilize the cellular mass. Also, forcing gas into a polymer melt or resin system will produce a "frothing" action resulting in a cellular polymer. Urea–formaldehyde, phenolics, polyesters, vinyl esters, acrylics, and some dispersion polymers are made cellular by mechanical techniques.

Syntactic polymers are produced by dispersing particles of rigid, flexible prepuffs, or microscopically small (0.03 mm) hollow balls of glass or plastics into a fluid polymer. Thermoplastic melts have been used but most are made from epoxies, polyesters, and urea–formaldehyde resin systems. The cellular material is a closed-cell struc-

Table 4-7. Selected Properties of Expanded Polymer

Polymer	Density (lb/ft³)	Tensile Strength (psi)	Compressive Strength (psi)	Thermal Conductivity (Btu·in./ft²·h·°F)	Service Temperature (°C)
ABS	35–55	2,000–4,000	2,300–3,700	0.58–2.1	80–85
Epoxy					
Rigid closed	5	50	90	0.26	200–260
Syntactic	14–20	40–80	125–150	0.49	250–300
Isocyanurate					
Semirigid	1.2–26	20–1,350	20–2,100	0.11–0.3	150
Phenolic	1–22	3–150	2–1,200	0.21–0.28	132–200
Syntactic	50–60	1,000	8,000–13,000	1.0	200–220
Polyether	1.4–17	20–500	20–600	0.10–0.3	100–150
Polyethylene					
Low-density	1.3–3	20–100	3–5	0.28–0.4	80–85
High-density	25–50	1,200	1,300	0.92	85–90
Polystyrene					
Beads	1–5	16–170	8–130	0.26–0.28	70–80
Extrusion	1.5–5	40–200	18–180	0.18–0.29	75–80
Polyurethane					
Rigid closed	1.3–70	15–8,000	15–15,000	0.11–0.57	140
Flexible	0.9–20	8–1,360	0.2–100	0.2–2	140
Polyvinyl chloride					
Flexible open	10–50	10–200	0.5–40	0.24–0.28	100–135
Flexible closed	4–11	50–150	0.5–40	0.24–0.28	100–135
Silicone					
Flexible closed	17–31	33–150	10–12	0.18–0.36	250–350
Urea–formaldehyde	0.8–1.2	2–10	5–20	0.18–0.2	100–150

ture. Major applications are for tooling, noise alleviation, and thermal insulation.

Sintering is a process of pressing particles of a polymer together. Polytetrafluoroethylene, polyamides, and other specially filled plastics are sintered into a microporous structure. The particles are compressed together but interconnected voids remain.

Leaching is a technique of dispersing soluble particles into the melt or resin mix. The open-cell structure of cellulose sponges is the most familiar polymer produced by the leaching process. Solvents are used to dissolve the salt crystals or plastic pieces dispersed in the polymer, leaving a porous, spongelike structure.

Coating Processes

The concept of coating as an important processing technique is often overlooked. Coating techniques encompass a broad range of materials, methods, and endless applications, and it is a highly developed technology. To be classed as coating, the polymer material must remain on the substrate. Coating processes are often confused with some casting processes, perhaps because similar equipment is used.

We coat various substrates for a multitude of reasons. Protection from the environments of corrosive exposure, abrasion, chemicals, vapors, solvents, or weathering are familiar examples. Some coatings are used for electrical insulation. In addition, coatings are used to add beauty and serviceability to the substrate; for example, a coating on a paper substrate can add beauty, strength, and resistance to scuffs, moisture, and soil and can provide a sealing system for making a package. The coatings on cars, houses, machinery, furniture, and fingernails are most familiar.

Most composite examples of coatings are modifications of laminating and reinforcing techniques. Coatings are generally considered to be a thin layer (or layers) of polymeric material covering a substrate. Protective coatings for large metal tanks are often accomplished by hand lay-up or spray-up methods. Containers that are to be placed underground must be protected from corrosion and electrolytic action. Chemical storage tanks may require protective coatings on the interior. In both examples, the primary reason for applying a layer of reinforcing material and matrix is to provide a protective coating: Improvement in strength is a secondary function.

In most coating operations it is important to select coating materials that are reasonably close to the thermal expansion of the substrate. Temperature gradients can cause stress and damage the usefulness of the coating. Most polymers expand at greater rates than other substrates. Selecting ductile or flexible coating materials or stabilizing them with reinforcements and fillers may be considered.

There are 12 broad (and sometimes overlapping) coating processes: (1) hand lay-up, (2) spray, (3) dip, (4) extrusion, (5) calender, (6) powder, (7) fluidized-bed, (8) electrostatic-bed, (9) transfer, (10) metal, (11) roller, and (12) brush. All are techniques for placing a polymer coating on a substrate.

Hand Lay-Up Coating

A hand lay-up coating is generally a protective layer of reinforcing material and a matrix of epoxy or polyester. As previously described, hand lay-up coatings are used on metal and wood tanks, pipes, ductwork, and vents to prevent corrosion.

Spray Coating

A spray coating is delivered to the substrate as a dispersion, solvent solution, or dry or molten powder. Variations differ in the method of atomizing the polymer. Air or compressed gas has been a conventional method for applying coatings on furniture, vehicles, and machinery. Polymer solutions may be atomized by hydraulic (airless) pressure. Flame coating methods use polymer powders and a special applicator spray gun to place molten coating on the substrate. Hot gases melt the powder as it passes through the gun and onto the substrate.

Dip Coating

Dip coating is a process where the substrate is simply dipped into a liquid-phase polymer. Liquid dispersion or solvent mixtures are most common, with PVC being widely used. Parts are heated and passed through the plastisol dispersion. The heat causes the dispersions to melt and stick to the hot surface. The coated part is then heated to cure and completely fuse the particles together. Organosols are solvent mixtures. Parts are dipped and the solvent is allowed to evaporate. The technique is sometimes called cold dipping

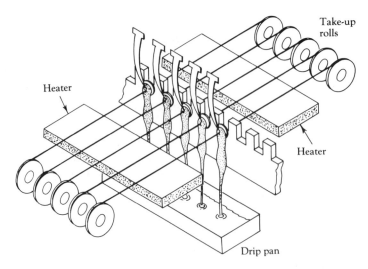

Figure 4-46. This modified dipping process can apply a plastisol coating to wire, cable, woven cord, or tubing at very high speeds. (BF Goodrich Chemical Division)

because the substrate is not heated. Dip coating techniques are illustrated in Figure 4-46.

Wires, cables, woven cords, tubing, tool handles, and dish-drainer racks are commonly dip coated. Objects are limited only by the size of the dipping tank and the heating method for plastisols.

Strippable coatings are sometimes included as a dip coating technique. Ethyl cellulose has been used to coat and protect parts for shipment. The parts are simply dipped into the heated polymer and removed for cooling. By definition these are more correctly called temporary coatings. Strippable coatings are sometimes used as masking films in painting or electroplating.

Extrusion Coating

Extrusion coating is a technique where the hot melt from the extruder is placed on (paper) or around (wire) the substrate. Several of the coating techniques result in a thin "laminated" composite. (See Laminating.) These materials are considered as coating techniques because the primary objective of the coating is to provide protection. Plastics do increase wet strength, abrasion resistance, and impact strength of paper and cardboard.

PE, PP, EVA, PET, PVC, PA, and other polymers are used in extrusion coating of various substrates.

Continuous web coatings are combined with the hot melt, forming a lamination of one or several layers. Paper, polymers, foils, and woven and nonwoven fabrics may be given a coating. Coatings provide moisture vapor barriers, gas barriers, liquid barriers, and heat-sealable surfaces. Polyethylene coating on paperboard milk cartons is a familiar example of liquid barrier and heat-sealable coatings. Many hybrid polymers and copolymers are combined in selective layering of polymer coatings and substrates.

Coextrusion of two or more polymers through the same manifold or multimanifold system may be used to combine with the substrate as the material exits the die slot. In all web extrusion coating operations, pressure rollers are used to press the hot melt against the substrate. These rollers held to determine the texture and thickness of the coating on the substrate. Adhesion is one of the most critical factors in the operation: Some substrates must be prepared with corona discharge, flame, plasma, ozone, or chemical treatments to assure good adhesion.

Cable coating extrusion is accomplished as wire is pulled through a manifold and die. The wire may be coated (encased) with solid or cellular polymers.

Calender Coating

Calender coating is a process where hot rollers are used to press the polymer melt against the substrate. Paper, polymers, and textiles are calender coated. Pressure-sensitive and heat-reactive hot melt polymers are commonly used as adhesives between multilayered products.

Powder Coating

Powder coating is a process of causing powdered polymers to strike, adhere, and coat a heated substrate. Some techniques require additional heating after coating to assure complete fusion. PE, EP, PA, CAB, PP, PU, ACS, PVC, DAP, AN, and PMMA are made into powder (solventless) formulations for various powder coating techniques.

One of the most effective methods for production line coating is electrostatic. Liquid or powdered polymers are atomized by high-speed electrostatic disks, electrostatic guns, air guns, or airless guns. The positively charged particles are then attracted to the negatively charged work piece. Polarity may be reversed for some operations. The electrostatic attraction causes most of the particles to cover all surfaces of the substrate. The technique may reduce overspray, improve finish quality, provide uniform coatings, and reduce waste over conventional air spray systems. Substrates may require special preparation by applying primers or other sensitizing formulations on the surface. Paper and plastics (nonconductive) may require that conductive agents be placed into the substrate formulation or that conductive primers be applied. Some metallic parts are heated and coated by the attraction of atomized, charged powder particulates. These powders melt and fuse together as they touch the hot substrate. If the parts are not preheated, they must be placed in the fusing or curing oven before the powdered particles lose their electrical charge and fall from the part.

Fluidized-Bed Coating

Fluidized-bed coating uses a special tank or container full of powder. The powder is made "fluid" by atomizing the powder with low air pressure forced through a porous membrane in the bottom of the tank. As the heated substrate is lowered into the fluidized powder (foglike), particles strike and adhere to the hot substrate. The part may be removed for cooling or returned to a heating chamber where the coating completes the fusing or curing phase.

A variation to fluidized-bed coating is fluidized-bed spray coating. The powder is simply atomized from a spray gun or head onto the heated part. Overspray is collected and reused.

Electrostatic-Bed Coating

Electrostatic-bed coating is a technique of atomizing a fine cloud of negatively charge polymer which is attracted to the positively charged substrate. Parts are then removed to the curing or fusing chamber. Parts may not need preheating if they retain an electrostatic charge to hold the powder on the substrate during cure.

Helmets, screens, gun stocks, pipes, skate boards, fencing, golf balls, typewriter cases, steering wheels, toilet seats, washing machines, furniture, dishwashers, and many small appliances are electrostatically coated.

Transfer Coating

Transfer coating begins by placing a skin coat on a release film or paper. After gel occurs, a second coat called the tie coat is applied to the skin coat. A textile or other substrate is then forced against this wet tie coat by nip rollers. The coated material is then dried in an oven and passed over cooling rollers. The release paper is then stripped from the textile. Several materials may be bonded together including cellular polymers with a film and/or fabric coating.

PU (30% solid) and PVC (100% solid) are the most commonly used polymers. Fabrics are coated for protection in the manufacture of awnings, footwear, upholstery, and fashion apparel.

Metal Coating

Metal coating is a technique of placing a thin metallic coating on a substrate. One of the simplest methods is to adhere metal foils to the polymer substrate. Complex or irregular shaped parts are difficult to coat by this method.

Electroplating is done on a number of polymers. Careful cleaning and etching of the surface are usually required to ensure adhesion of the metal ions.

Vacuum metallizing is a technique used to coat polymers with vaporized metal in a vacuum chamber. After the parts are cleaned and etched as needed, they are placed in a vacuum chamber. Aluminum, gold, silver, or zinc may be melted until it vaporizes and coats everything the vapor touches. By alternately evaporating two metals, it is possible to create a chrome/copper/chrome or stainless/copper/stainless laminate. The process is sometimes called **laminated vapor plating**.

Sputter coatings of metals or refractories may be used on some polymer surfaces. Atoms of molten metal are ejected from a source and strike the polymer surface. The process is done in a high-vacuum chamber. A clear protective coating of polyurethane, acrylate, or cellulosic is applied after sputtering. Magnetron discharge of molten metals is used to coat plumbing fixtures, films, and light reflectors.

Metal coatings are used to produce mirrorlike finishes on parts.

Most are decorative but some are used to provide an electrically conductive path or added heat deflection in insulation applications.

Roller Coating

Roller coating refers to a number of roller configurations and methods of placing a coating on a substrate. All production techniques involve a continuous web operation. Air knife, kiss, squeeze-roll, gravure, and reverse roll are all descriptive of roll coating variations. All variations are used to place a thin, even coating on a substrate.

Brush Coating

Brush coating techniques are familiar to every homeowner. Polymer coatings may be applied by hand with brush, pads, or other applicator aids. Solvent and solventless polymer coatings are used. Solvent coatings need only air drying or heating to cure, while solventless coatings usually require catalysts or postcuring. Polyester and epoxy are the most familiar examples. Brush coating finish quality depends largely on the skill of the applicator. Most operations are for repair of damaged substrates, coatings on short-run, experimental pieces, or for jobs requiring on-site fabrication.

Fabrication Processes

We are familiar with fabrication or assembly techniques and processes used in the wood and metal industries. Many of these familiar techniques are used in the fabrication of polymers and composite materials. The word **fabrication** is used to describe methods of attachment (assembly), not operations involved in the production of composite parts. Most may be considered secondary operations, which fall into one of four broad fabrication methods: (1) adhesion, (2) cohesion, (3) mechanical fastening, and (4) friction fitting.

Fabrication techniques are important in reducing both structural weight and costs in composite components. These techniques must be considered when design parameters are viewed. Strength, service environment, applied stresses, life of the joint, and compatibility with other materials in the assembly must be carefully considered. (See Basic Design Practices.)

Adhesion

Adhesion has been defined as a state where two surfaces are held (adhered) together by interface or interlocking forces. This bonding does not cause mingling of molecules between surfaces. Wood and paper are commonly held together by physical or mechanical adhesion. This usually involves van der Waals forces and simple mechanical interlocking on the surfaces.

Adhesive bonding is a common, efficient, economical, and durable method for fabricating components. Although there are many texts on the subject, not all adhesive technologists agree on how an adhesive functions or what should be included as an adhesive. In this text, an adhesive includes all substances used to hold materials together by surface attachment. These adhesives are substances (thin layers) used to spread stresses from one substrate to another. This would include solder, frozen water, cements, waxes, natural resins, gums, or other materials that satisfy this definition. No distinction is made between permanent structural adhesives (sandwich wing composite), temporary or demountable adhesives (masking tape), and nonstructural adhesives (floor coverings).

Adhesives come in a number of available forms: powders, films, tapes, dispersions, hot melts, pastes, liquids, and two-part components. All must be made into a viscous liquid at some point in the fabrication process in order to wet the substrate surface for high-quality, durable adhesion.

Most substrates require some form of surface preparation to ensure dependable, high-quality bonds. Mold releases, machining oils, and oxides must be removed. Some substrates may require only solvent wiping or abrasion cleaning. Many polymers and reactive metals must be cleaned by chemical treatments such as oxidizing flame, plasma, etching, and chemical baths. Specific cleaning and surface treatment procedures are described in several references. (See Bibliography.)

Selecting the correct adhesive system is very important. Some solvents in the adhesive may attack or cause chemical stress cracking. Assembly methods, adhesive cost, bonding cycle, and ultimate extremes of intended application must be involved in the selection. It is wise to contact the adhesive manufacturer for recommendations, testing results, and intended application.

There are normally several adhesive materials from which to choose when bonding any two surfaces: polymer-to-polymer, polymer-to-metal, nonpolymer-to-polymer, and so on. **Elastomeric**-type adhesives are popular where joint flexibility is desirable. **Thermoplastic**-type adhesives are generally solvent based or hot melts. Cyanoacrylate cures through polymerization. **Thermosetting resin** adhesives require chemical mixing, heat curing, or other polymerization mechanisms to activate the curing system. There are a number of miscellaneous adhesives, including pressure-activated tapes, water-based cements, and caulking.

There are a variety of techniques for applying adhesives, most depending on adhesive type, factors relating to automated production, accessibility of joint, joint design, surface area, joint quality, personnel safety, and labor costs.

Applications for adhesives encompass a broad field. Automotive, textile, aerospace, construction, packaging, and electronics industries continue to show significant application gains.

Adhesives are used to replace mechanical fasteners (screws, threads, etc.) and welds. Vinyl car tops, fire wall insulation, headliners, structural reinforcing panels, furniture assemblies, decorative strips, paperboard products, shoe soles, wall papers, stitchless seamed clothing and many more applications use adhesive bonding.

Cohesion

The chemical adhesion that holds a material together is called **cohesion**. Cohesion results from the strong valence attraction between substrates that are caused to flow together such as when metal surfaces melt and flow together during welding. In cohesive bonding, there is an intermingling of molecules between substrates. Cohesive bonding methods may include (1) cement, (2) spin, (3) hot-gas weld, (4) heated-tool, (5) impulse, (6) dielectric, (7) ultrasonic, (8) electromagnetic, and (9) co-curing.

Cement

Cement bonding refers to solvent or solvent mixes used to dissolve the substrates and fuse them together. Solvent evaporates from the joint region, resulting in chemical stress cracking and decreased joint strength. Dope cements, which are viscous materials, are sometimes called laminating cements. Careful consideration must be given to the proper selection of solvent for each polymer. Solvent cementing is usually done on like thermoplastic substrates. No thermosetting and only selected thermoplastic materials are solvent cemented.

Spin

Spin bonding (spin welding) is a process that uses the heat from friction to melt substrates together. Circular thermoplastic parts are bonded by this method. Part joints or filler rods may be spun to create the weld joint. Once the melt cools the bond is completed. Assembly of plastic bottles, tops, and other containers are common examples. Rapid back-and-forth movement, ranging from 100 to 300 Hz in frequency and sometimes called vibration welding, may be used in some part joint designs. Most thermoplastic composites can be bonded, including dissimilar polymers with compatible melt temperatures.

Hot-Gas Weld

Hot-gas weld bonding is a flameless process. Heated gas (usually nitrogen) is used to melt the substrates and/or filler material. The process is similar to flame welding of metals. The electrically heated gas and gas pressure can be controlled for proper welding parameters. Large pipes, ducts, and tanks may be fabricated by hot-gas weld bonding.

Heated-Tool

Heated-tool bonding (fusion welding) is a relatively simple cohesive method. Continuously heated dies, tools, or plates are used to melt and fuse the like materials together. In some operations, the heated tool simply presses the heated materials together until melt occurs at the joint. Joints may also be heated by hot dies. The dies are removed and the polymers are quickly brought together while in the molten phase. Nearly all polymers may be made heat sealable by applying a layer or coating of material that is heat sealable. Joining of pipes, films, and sheet edges are examples of the process.

Impulse

Impulse bonding is similar to heated-tool bonding except that the tooling is not continuously heated. Heating tools are quickly heated for a predetermined length of time. The heated joint is held under pressure until the tool and the melt cool. Plastic packages are commonly impulse bonded.

Dielectric

Dielectric (radio-frequency) bonding involves the use of high-frequency (10–200 MHz) waves to cause rapid oscillation of the polymer molecules. Frictional heat from this movement causes the polymer to become molten. Only those polymers with high dielectric loss characteristics (dissipation factor) may be joined by this method. Polyethylene, polystyrene, and fluoroplastics have a low dissipation factor. Thus, these materials, including papers and fibers, can be dielectrically sealed by using bonding agents that are radio-frequency sensitive.

Ultrasonic

Ultrasonic bonding uses high-frequency (20–40 kHz) mechanical vibrations directed by the bonding tool called a horn. These frequencies cause the polymer molecules to vibrate and the generated frictional heat causes the polymer to become molten. The designs may be used for spot bonding or staking (rivet heating) of plastic studs. A spurlike tool is used to make stitchlike bonds in some fabrics and films. Ultrasonics is sometimes used to aid in the insertion of metallic fasteners or as the energy source for curing some adhesives.

Electromagnetic

Electromagnetic bonding uses micron-sized particles of iron oxide, stainless steel, ceramic, ferrites, and graphite which respond to the radio-frequency (3–40 MHz) magnetic field. These (opaque) powders or inserts must be molded in the polymer matrix and remain in the final weld. As the high frequency field passes through the magnetic materials, they are induced to become hot. The surrounding polymer melts, forming the bond. The induction coil must be located as close to the joint as possible if rapid bonds are to be made. Nonmetallic tooling must be used for alignment. Reinforced, coextruded layers, sandwich, and nonthermoplastic (thermosets, paper, foils) coated composites may be bonded. Some are bonded by incorporating magnetic particles in hot melts, liquid adhesives, or films at the joint area.

Structural bonds requiring a hermetic seal are possible in automobile panels, solar panels, pipes, and containers.

Co-Curing

Co-curing bonding may be considered an adhesive or cohesive bond depending on the similarity of materials. Co-curing is an important bonding technique. It refers to the technique of bonding together two or more composite layers using B-stage matrix resins (or adhesives). The faces (substrates) are co-cured into one cohesive piece by the application of some form of energy. If dissimilar materials such as epoxy–carbon and an aluminum honeycomb core are laminated by this technique, an adhesive bond is the result. Only when there is an intermingling of molecules is there a cohesive bond. Composite components of epoxy–glass, epoxy–aramid, epoxy–glass layers may be considered as cohesively bonded. Although this latter example fits the cited definition for fabrication, many would consider this a production operation even if it were a secondary operation of attachment.

Mechanical Fastening

There are numerous metallic and plastic fasteners used for high-strength assembly or where designs cannot compensate for the weakness of adhesives (high peel stresses). If repeated disassembly is anticipated, mechanical fasteners may be used. Molded-in inserts, press-in inserts, or thread-forming inserts are commonly used if periodic die assembly is required. Expansion inserts, threaded clips, nuts, bolts, rivets, self-tapping screws, and other fasteners are also used. The use of mechanical fasteners must be considered carefully because of stress concentrations. (See Basic Design Practice.)

Friction Fitting

Friction fitting is a term used to describe a number of pressure-tight joints for permanent or temporary assemblies. Press fits, shrink fits, and snap fits are variations of friction fitting designs. The difference between a snap fit and a press fit is the undercut and force required for assembly. The distinction between a press fit and a snap fit (dual cantilever) is illustrated in Figure 4-47. Shrink fitting refers to placing (cold) inserts into the hot molded plastics before the part has cooled. It also refers to plastics that shrink by a property called

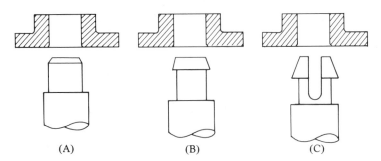

Figure 4-47. (A) Press fit with interference stress and critical tolerances. (B) Hollow-cylinder snap with tight tolerances but less interference. (C) Dual cantilever snap with less critical tolerances.

memory. Electronic components or packages may be covered by heating a shrinkable tube or wrapping. Several boxes may be held together on a pallet by shrink fit wrapping or bag.

Radiation Processes

Radiation processing may include ionizing and nonionizing systems. Ionizing radiation causes (ionization) atoms and molecules to become highly reactive. If carefully controlled, this may result in major improvements in physical properties. Radiation processes have found acceptance along with chemical and thermal processing.

Nonionizing radiation may include induction, dielectric, microwave, ultraviolet, and infrared sources. Many of these sources may (with sufficient energy) cause radiation damage and break molecular bonds. At low energy levels, they are generally used to process by heating, drying, and curing selected adhesives or thin film layers.

Electron beam accelerators are the primary ionizing source for production processing. Radioactive materials such as cobalt-60 are more difficult to control (shut off) and direct. Accelerator voltage and electron beam current are more easily controlled or delivered to the work area. Irradiation processing implies that a controlled or directed treatment of energy is being transferred from the radiation source to the polymeric material.

The effects of radiation on polymers may include (1) damage, (2) property improvements, and (3) polymerization and grafting.

Damage

All radiation may be harmful to polymers, with some being more easily damaged than others. When molecular bonds are broken, scission, disassociation, displacement of atoms, or formation of radicals may result. This in turn can lead to lower molecular weight, cross-linking, branching, polymerization, or oxidation. Cross-linkage in elastomers may be considered degradation. Fillers, chemical additives, and cover pigmentation may prevent deep penetration of damaging radiation. The effects of radiation on selected polymers are shown in Tables 4-8, 4-9, and 4-10.

Table 4-8. Effects of Radiation on Selected Thermoplastic Polymers

Polymer	Radiation Stability	Radiation Dose to Produce Significant Damage (Mrads)
ABS	Good	100
Acetals	Poor	1–2
Acrylics		
PMMA	Fair	5
Others	Fair	10
Amides		
Aliphatic	Fair	50–100
Aromatic	Excellent	1000
Cellulosics	Fair	20
Fluoroplastics		
PTFE	Poor	1
PCTFE	Fair	10–20
FEP	Fair	20
PFV, PETFE, PECTFE	Good	100
Polycarbonate	Good	100+
Polyesters (aromatic)	Good	100
Polyolefins		
Polyethylene	Good	100
Polypropylene	Fair	10
Polymethylpentene	Good	30–50
Copolymers	Good	50
Polystyrene	Excellent	1000
Copolymer	Good	100–500
Polysulfones	Excellent	1000
Polyvinyls		
PVC	Good	50–100
Copolymer	Fair	10–40

Table 4-9. Radiation Sensitivity of Selected Thermoset Polymers

Material	Radiation Resistance	Radiation Dose for Significant Damage (Mrads)
Epoxies	Excellent	100–10,000
Phenol (or urea) formaldehyde	Good	500
Polyesters (unsaturated)	Good	1,000
Polyimides	Excellent	100–10,000
Polyurethanes	Excellent	1,000 +

Property Improvements

Controlled irradiation may benefit some polymers. As a result of cross-linking, grafting, and branching, thermoplastic materials can have marked improvements in temperature and physical properties. The shrink-memory property of heat-shrinkable plastics and the elimination or reduction of residual monomers in packaged products are beneficial. Some surface treatments improve weather, penetration, and static resistances. Cold sterilization of packaging and surgical supplies are important processing aids.

Polymerization and Grafting

Cross-linking and grafting may be done on previously shaped parts. Irradiation may cause two or more elastomers to attach to the

Table 4-10. Radiation Sensitivity of Selected Elastomers

Material (Rubbers)	Radiation Threshold Damage (Mrads)	Dosage 25% Damage (Mrads)
Polyacrylic	2–4	10–15
Butyl	2–3	10
Chlorosulfonated polyethylene	—	30
EPDM (ethylene–propylene–diene)	10	100+
Fluoroelastomer	5	50–70
Natural	10	100–200
Nitrile	—	100
Polychloroprene (Neoprene)	6	50
Silicones	—	50–100
Styrene–butadiene	6–8	100+
Urethane	20	600–800

Table 4-11. Industrial Applications for Electron Beam Processing

Product	Product Improvements and Process Advantages	Process
Wire and cable insulation, plastic insulating tubing, plastic packaging film	Shrinkability; impact strength; cut-through, heat, solvent, stress-cracking resistance; low dielectric losses	Cross-linking, vulcanization
Foamed polyethylene	Compression and tensile strength; reduced elongation	Cross-linking, vulcanization
Natural and synthetic rubber	High-temperature stability; abrasion resistance; cold vulcanization; elimination of vulcanizing agents	Cross-linking, vulcanization
Adhesives Pressure sensitive Flock Laminate	Increased bonding; chemical, chipping, abrasion, weathering resistance; elimination of solvent; 100%	Curing, polymerization
Coatings, paints, and inks on Woods Metals Plastics	convertibility of coating; high-speed cure, flexibility in handling techniques; low energy consumption; room-temperature cure; no limitation on colors	Curing, polymerization
Wood and organic impregnates	Mar, scratch, abrasion, warping, swelling, weathering resistance; dimensional stability; surface uniformity; upgrading of softwoods	Curing, polymerization
Cellulose	Enhanced chemical combination	Depolymerization
Textiles and textile fibers	Soil-release; crease, shrink, weathering resistance; improved dyeability; static dissipation; thermal stability	Grafting
Film and paper	Surface adhesion; improved wettability	Grafting

molecular chain (grafting) without the addition of chemical catalysts. Fillers and reinforcements have little effect.

Polymerization and cross-linking are important irradiation processing techniques used to cure polymer coatings, adhesives, or monomer layers. Typical radiation doses (Mrads) for cross-linkable polymers may range from 20 to 30 for polyethylene, 5 to 8 for PVC, 8 to 16 for polyvinylidene fluoride, 10 to 15 for ethylene vinyl acetate, and 6 to 10 for ethylene chlorotrifluoroethylene. Cross-linking of wire insulation, elastomers (vulcanization), and plastic components are used to improve stress-cracking, abrasion, chemical, and deformation resistances. Table 4-11 lists a few of the industrial applications for electron beam processing.

The processing advantages for composite components and structures have been used to cure B-stage resins with less chemical-related stresses and physical damage to the reinforcing additives. Aramid fibers are susceptible to some degradation but are generally superior to that of most polyamides and other organic fibers.

Chapter 5. What Are the Mold and Tooling Parameters?

Introduction

To deal with the broad, multifaceted concepts of composite tooling, the reader should review the descriptions of polymer families (Chapters 1–3) and individual processing techniques (Chapter 4). This chapter contains information about properties and design parameters that affect moldability. Design standards, specifications, and practices are discussed in Chapter 6.

CAMM

Today, tool planning involves a number of interactive inputs from several sources, systems, or personnel. Computer-assisted mold-making (CAMM) is a sophisticated tool (instrument) that a mold-maker can use to increase productivity. The composite industry must supply competitively priced components and structures with high-quality surfaces, complex configurations, close tolerances, and reliable designs. Quality is now the buzzword for countering foreign competition. (See Figure 5-1.)

CIM

Computer-integrated manufacturing is a somewhat loose term indicating that all aspects of manufacturing are supported and perhaps controlled (integrated) by one or more computers.

The introduction of computers for tool plan design and produc-

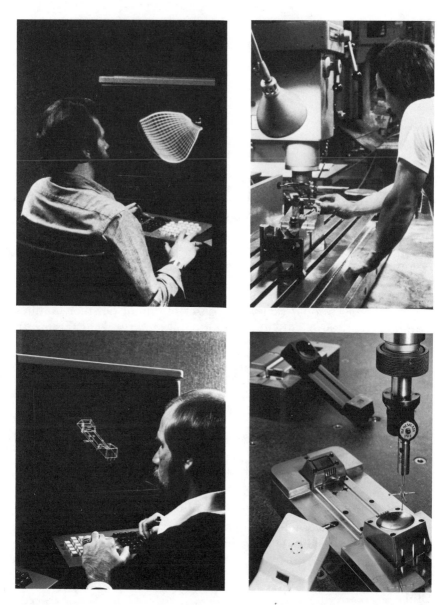

Figure 5-1. Computers are used in design, mold manufacture, mold flow analysis, and quality control. Gone are the days when all a moldmaker had to do was produce a mold to match a blueprint. (Minco Tool and Mold, Inc.)

tion has the greatest effect on increasing efficiency. Moldmaking is a highly labor-intensive industry and custom moldmaking shops offering special service such as plating, polishing, heat-treating, engraving, or matching molds continue to compete for a limited supply of skilled designers, engineers, and toolmakers. Many of the current skills were learned through on-the-job training or in apprenticeship programs. With the changing design parameters and demands for increased productivity there is little margin for trial-and-error methods or prototype models.

One way to meet the challenge of skilled personnel shortages and increased productivity demands is through the use of computer-aided design (also drafting) (CAD), computer-aided manufacturing (CAM), and computer-aided engineering (CAE) systems.

Like any system, a computer system is only as useful as its weakest link. To gain full benefit from a CAD/CAM system, all information (data base) must be shared between the drafting, designing, engineering, moldmaking, and manufacturing departments. Costly redefinition, interpretation, and communication errors are thus eliminated. Since each area has access to the data, product quality, reliability, and productivity are increased.

Local-area network (LAN) standards for computer-to-computer communications are the keys to making CIM work. This can mean networking for a few local electronic devices to communications with devices anywhere in the country. General Motors has funded the development of Manufacturing Automation Protocol (MAP), whose aim is a set of national standards for plant automation and data communications. Boeing has developed a network protocol, called Technical and Office Protocol (TOP), for plant offices and laboratories. Many firms and organizations are spearheading the drive for universal LAN standards, including General Motors, Boeing, the International Organization for Standardization (ISO), the American National Standards Institute (ANSI), the Institute of Electrical and Electronics Engineers (IEEE), and the National Bureau of Standards.

The lack of standardization among machines, controls, and auxiliary elements has complicated the shift into computer technology for some moldmakers. Most cannot simply discard existing equipment and must consequently contend with diverse programming, controls, and interface modules.

CAD

Computer-aided design is a term for design activity that is supported by a computer on an interactive basis.

CAD systems are used to create and draw mold and tooling designs. Modifications made in the illustration are automatically changed in all views. Views can be rotated, color enhanced, or enlarged, and details can be selectively erased or added. Drafting time and the change in tooling parameters often represent a significant percentage of the total cost of a part. The drafting and designing departments can use CAD systems that automatically dimension the drawing and allow for material shrinkage. Any needed modifications in the drawings can quickly be accomplished and stored in the data base for everyone to use. Commonly used or preparatory patterns, symbols, or standard mold bases and components are also used in interactive systems.

Once the initial design and drawing are completed, copies can be made on hard copy devices or multicolor plotters. The designer may simply define the tooling parameters by answering a series of questions. The program will then assist in defining (selecting) the mold base and in suggesting where other components such as ejector pins, sleeves, locating ring, return pins, pullers, and many other parts are to be located.

CAM

Computer-aided manufacturing is a term used to indicate any activity supported and perhaps controlled by a computer. The final tooling configurations and manufacturing techniques must then be determined.

CAM systems are used to determine how the core, cavity, runners, cooling system, system finish, and complex shapes are to be machined, so that many problems can be eliminated before any material is cut. The cutting tool path, part geometry, feeds, and inherent equipment limitations (backlash, loading, tool wear) can be viewed by the user. The tool movement can be verified and the program can be loaded directly on the machine controller or placed on tapes. Jigs and fixtures can also be generated to depict hold down fixtures needed in the machining of the mold base, tooling, or secondary operations such as painting, finishing, or packaging. It has been estimated that nearly 80% of the moldmakers' time is devoted to setting up the machine tool: Only 20% is actually spent cutting material. CAD/CAM systems substantially reduce setup, lead, and machining times.

In a CAD/CAM system, the design information generated on a CAD system is readily usable by a CAM system.

CAE

Computer-aided engineering is not a synonym for CAD/CAM, but a step up from CAD. It may include finite-element and other design analysis, system modeling, and simulated structural testing.

CAE is a broad spectrum of analytical engineering functions that focus on detailed designs, drawings, NC tape preparations, selection of proper materials, cavity filling, cooling, and structural analysis of the tooling and composite part. Computer models can be viewed for flow analysis. Runners, gates, or orientation of the composite can be predicted by using analysis and models provided by the data base. The effects of moving gates, changing runner dimensions, percentage of reinforcements, matrix, or other parameters may be viewed by the user for optimum configurations. This minimizes material waste. Material selection, cycle times, part cost, and tooling approvals are quickly made. The approved design may then be sent by tape or other communication linkage to the tooling department. (See Figure 5-2.)

CAD/CAM/CAE can save time and costs in virtually every phase of the composite product development process from concept to completion. The most important benefit of computers in engineering is the improvement in design, labor productivity, market share, capital productivity, innovation, quality products, and profitability. None of these benefits can be attained if a company allocates access to the data base for only a small cadre of personnel. Designers, engineers, manufacturers, moldmakers, and others must be integrated into or have access to the system data base. It does little good if a designer uses mathematical modeling of the cooling process for a mold and determines that imbalanced cooling will cause unacceptable part warpage if that information is not shared with the tooling department. The benefits of CAD/CAM/CAE are illustrated in Figure 5-3.

Types of Tooling

Selection of the tooling materials and the actual tooling costs vary greatly. Some selection is based on the tool cost, physical size

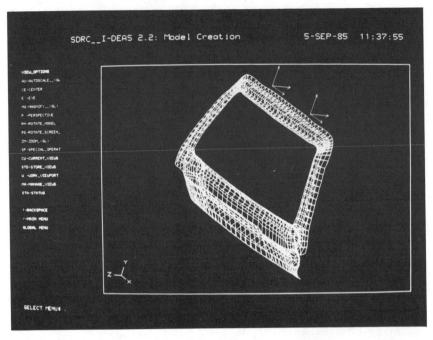

Figure 5-2. From information displayed on the computer, personnel can use finite-element analysis to test a new composite automotive liftgate design. (Owens–Corning Fiberglas Corporation)

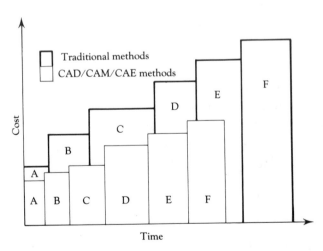

Figure 5-3. Benefits of CAD/CAM/CAE. A = concept development, B = drafting/revisions, C = tool designing/redesigning, D = cost and performance evaluation, E = tooling, F = part production.

Table 5-1. General Types of Composite Tooling

Tool Classification	Number of Parts	Tooling Materials
Prototype	1–10	Plaster, wood, reinforced plasters
Temporary	10–100	Faced plasters, reinforced plasters, faced plastics, backed metal deposition, cast and machined soft metals
Short run	100–1000	Soft-metal steels
Production	>1000	Steel, soft metals for some processes

of the part, part finish (tool surface), production technique, fiber orientation, additives, compatibility of thermal expansion (tooling and composite), curing method, matrix, and the size of the production run (service life). Complexity of design, rapid heat transfer, number of duplicate cavities, and required tolerances are also considered. (No tooling material will satisfy all these requirements.) Storage, tool life, mold maintenance, and machining are all part of the tooling cost.

There are only four general types of composite tooling: (1) permanent, long-run, (2) short-run, (3) temporary, and (4) prototype. As a rule, metals are selected for long-run, permanent tooling and gypsum plasters, plastics, and wood for short-run, temporary, prototype work. (See Table 5-1.)

Advantages and disadvantages of selected tooling materials for composites are shown in Table 5-2.

Table 5-2. Advantages and Disadvantages of Selected Tooling Materials for Composites

Tool Material	Advantages	Disadvantages
Aluminum	Low cost; good heat transfer; easily machined; corrosion resistant; lightweight; rustproof	Porosity; softness; galling; thermal expansion; easily damaged; limited runs
Copper alloys (brass, bronzes, beryllium)	Easily machined; good surface detail; high thermal conductivity; rustproof	Softness; copper may inhibit cure; attacked by some acids; easily damaged; limited runs

Continued

Table 5-2. Continued

Tool Material	Advantages	Disadvantages
Plaster	Low cost; easily shaped; good dimensional stability; rustproof	Porosity; softness; poor thermal conductivity; easily damaged; limited runs; limited thermal and strength ranges
Polymers (laminated, reinforced, filled)	Low cost; easily fabricated; thermal expansion similar to many composites; lightweight; rustproof; economical large designs; fewer parts	Limited design; poor thermal conductivity; limited dimensional stability; limited runs; limited thermal range
Steel	Most durable; high thermal resistance; strong and wear resistant; thermal conductivity	Most expensive tooling; difficult machining; size limitations; many parts; rusts; heavy
Wood	Low cost; easily machined; lightweight; rustproof	Porosity; poor dimensional stability; soft; limited runs; poor thermal conductivity and resistance
Zinc alloys (lead, tin)	Low cost; easily machined; good thermal conductivity; good detail; rustproof	Soft; easily damaged; limited runs; limited thermal and strength ranges
Miscellaneous (salts, inflatables, wax, ceramics)	Low cost; reusable; designed with undercuts; easily fabricated; light; hard; thermal conductors	Soft; easily damaged; dimensionally unstable; damaged by high temperature or chemicals; poor thermal conductors

Metal Tooling

Metal tooling is essential for many operations in forming composite materials. Long production runs require reliable, durable tooling. Many processing techniques require that molds be made of materials with high heat conductivity, strength, and corrosion re-

Table 5-3. Thermal Properties of Selected Metals

Property	Be	Mg	Al	Ti	Fe	Cu
Melting temperature (°C)	1277	650	660	1668	1536	1083
Thermal conductivity (W/m·K)	146	153	222	171	75	393
Linear thermal expansion (μm/μm/°C)	0.29	0.68	0.59	0.21	0.30	0.41
Specific heat (J/kg·K)	1883	1046	900	519	460	385

sistance. The thermal properties of selected metals are shown in Table 5-3.

Aluminum

Aluminum 7075 is a popular, high-grade, heat-treated alloy that has a carefully controlled composition in the AlZnMgCu-type range. It has a thermal conductivity nearly four times greater than steel and a relative density of about 2.75. Aluminum has a linear thermal expansion similar to some composite laminates or about twice that of steel. Chemical corrosion will occur when using PVC and some blowing agents. Aluminum 7075 is machined on standard equipment, including lathes, EDM, borers, millers, punches, and engravers. Surface finish may be in the 3–5 μm range with a hardness of about 60 RB. Anodizing and plating improve surface hardness and prolong tooling life. Wear, damage, or machining errors may be repaired with appropriate TIG welding procedures.

Aluminum tooling (7075-T652) is used extensively in blow molding, bag molding, thermoforming, and prototype models or molds. Because aluminum lacks sufficient strength and hardness, pinch-off areas usually have steel inserts. The inherent softness of the metal allows for rapid wash and wear in injection mold prototypes. Long-run thermoforming molds and dies should be Teflon or hard coated.

Copper

Copper alloys are an alternative to aluminum. Copper–nickel (C71500) and copper–beryllium (C17200) are commonly used in blow molding. CuBe with a relative density of 8.09 is about three times heavier than aluminum. The greater strength and a hardness of about 65 RB are significant advantages for achieving long service life. Copper alloys are easily repaired by TIG welding. Tooling applications are similar to those for aluminum. Although they have considerable hardness, copper alloys lack the strength and wear resistance of steel in long production runs for injection molding of composite materials.

Zinc

Zinc alloys (AC41A) with CuAlMg compositions are popular pre-production or short-production-run molding materials. They are soft and pinch-off edges must be protected by inserts against excessive stress concentrations. The mechanical values of zinc alloys decrease significantly at temperatures of about 100°C. Because they are easily machined and have very good moldability by casting, zinc alloys are an important tooling material for selected composite molding and casting processes. Zinc (Zamak, Kirksite) has been used for long-run blow molding operations that exceed 500,000 components, depending on the complexity of design and the tolerance limits.

Most soft metals, aluminum, copper, and zinc alloys have enjoyed popularity as inexpensive tooling for many low-pressure molding and casting processes. Steam expanded, chemical foaming, blow molding, gravity casting, thermoforming, hand lay-up, spray-up, bag molding, and dip casting are the most familiar operations.

These metals are also used for a number of metal-deposition methods in moldmaking. Metal deposition is a process of depositing molten metal on a steel master mold or mandrel. The coating is then removed and backed for additional strength. The molds or mandrels are reused to produce numerous duplicate molds.

These metals are cast over master molds (hub, sometimes called hob) to form a cavity. The process shown in Figure 5-4 is sometimes called **hot hubbing**.

Electroforming is an electroplating process used to deposit a metal coating on master molds made of plastics, wax, or metal. The thin shell coating is then removed and used as a cavity for thermoforming, RIM, LIM, and various casting and blow molding processes. Sputter, flame spraying, and vacuum metallizing are used in a similar manner to produce mold cavities.

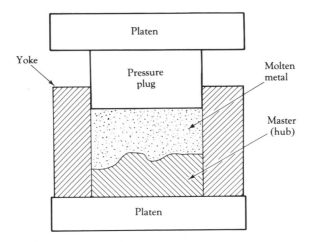

Figure 5-4. Hot hubbing a mold cavity.

Steel

Steel tooling is essential to the polymer composite industry. For long production runs with reliable finishes, product quality, wear resistance, and strength at elevated temperatures, steel is vital. Tool-makers have a great variety of steels from which to choose. Selection may be based on wear resistance, shock resistance, and ease of machining and treating the tooling. Some are preferred for machine parts, slides, guides, and bases, while others are used extensively as molding cavities. (See Figure 5-5.)

Tooling steels may be divided into the following groups: (1) non-alloyed (carbon steels), (2) alloyed, and (3) special-purpose steels.

The American Iron and Steel Institute (AISI), the Society of Automotive Engineers (SAE), and the American Society for Testing Materials have identification and designation systems for most compositions of steel. There are additional foreign and proprietary grades and designations. No attempt is made here to list numerous grades or distinguish formulations. Only a few steels important to the tool-maker are presented.

Nonalloyed Steels

Nonalloyed steels have varying amounts of carbon (up to about 2%). Some may consider carbon an alloy of steel, but it is an alloy that turns iron into steel.

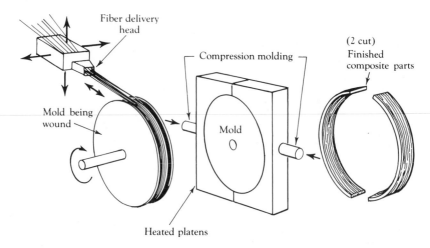

Figure 5-5. Steel tooling is used for the winding mandrel and compression molding of these composite parts.

Popular carbon grade steels used for machine bases, frames, and structural components are AISI 1020, 1025, 1030, 1040, and 1045. As a rule, tool steels with low carbon content have good impact resistance, while those with high carbon content have high abrasion resistance.

Alloyed Steels

Alloyed steels are carbon steels with limited amounts of alloy added (< 5%). Tool steels fit into this imprecise definition and include steels that may contain more alloying elements than alloyed steels.

Alloying elements are added to overcome many of the shortcomings of carbon steels. The general effect of alloying carbon steels is shown in Table 5-4. Many steels may be made more useful by any of a number of surface-hardening methods, such as carburizing and case hardening, cyaniding, carbonitriding, nitriding, flame, and induction hardening. Steels with alloys of aluminum, chromium nickel, and molybdenum are commonly nitrided. This surface treatment produces an extremely hard (70 RC) layer. Nitrided steels are used to withstand the abrasive wear of composite materials. In addition, they are corrosion resistant and stable at temperatures to 800°F (427°C).

Table 5-4. General Effects of Alloying Steels

Alloy Material	Percentage (%)	Effects
Aluminum	<2	Aids nitriding, deoxidizer
Chromium	0.3–4	Increases hardenability, wear, abrasion, and corrosion resistances
Lead	0.05–0.3	Improves machinability
Manganese	0.3–2	Increases hardenability, depth of hardness, and abrasion resistance
Molybdenum	0.1–0.5	Increases hardenability, hot strength, and creep strength
Nickel	0.3–5	Increases hardenability, toughness, annd corrosion resistance
Phosphorus	0.05–0.15	Increases machinability, hardenability, and corrosion resistance
Silicon	0.2–3	Increases hardenability, toughness, and corrosion resistance
Sulfur	0.07–0.3	Improves machinability with manganese
Vanadium	0.1–0.3	Increases hardenability, wear resistance, and toughness

To improve machinability, sulfur, lead, or manganese is sometimes added. Normalizing and annealing are also used to create desirable machining characteristics. Selenium is often added to stainless steel to improve machinability and resist work-hardening qualities.

Common steels for cold-work tool steel hubs are the D series. These tool steels have high carbon and chromium contents. The AISI type A series are cold-work air-hardened tool steels popular with all toolmakers. Types A2 and A6 are commonly used for transfer, compression, and master hub moldmaking. Fully hardened tool steels such as D2 and D3 have been used successfully for injection and compression molds. High-carbon and -chromium D3 steels have good wear resistance. Suitable molding cavities, back plates, knives, punches, and dies have been produced from the chromium–molybdenum steel 4140. This steel is tough and may be hardened to more than 30 RC.

Special-Purpose Steels

Special steels supplied in the solution-annealed condition are the machining steels for polymer molds. The soft nickel–martensite content allows these steels to be readily machined. Volume changes must be taken into consideration prior to machining and heat treatment. The AISI 18 MAR 250 and 300 are used. If composite materials are to be molded, hard-chrome plating or nitriding are recommended.

A special type of steel, which is tough, hard, and easy to machine, has been formulated for mold tooling. For normal stresses and wear resistance this mold steel is adequate. Wear resistance may be improved by chromium plating or nitriding. The chromium–manganese alloys P2 and P5 are the most commonly used in hubbing operations because of their very low carbon content. Mold cavities, holding blocks, dies, and other tooling are made of P20 steel. The AISI P20 and P21 have sufficient carbon and alloy content to harden to more than 35 RC.

Corrosion-resistant steels are used for processing chemically corrosive composites. Stainless steels of the AISI type 420 and 440C have been hardened to more than 50 RC for corrosion- and wear-resistant tooling.

Plaster Tooling

Plaster tooling has been a popular, low-cost tooling material for prototype creation and production of composite materials for many years. There are a number of high-strength gypsum plasters developed to meet the various needs of the moldmaker and processing technique. These materials have enough strength to produce prototype models, original master models, die models, transfer (take-off) tools, patterns, and die molds. (See Figure 5-6.) They may be further strengthened by adding reinforcing fibers. Metal bases and frames are preferred mountings because they provide a rigid, dimensionally stable, moisture-resistant supporting medium for the plaster.

Loft template tooling methods are widely used to produce tooling designs. In this technique templates are mounted together and secured to the base support. Expanded metal or other reinforcing or filler support materials may be placed between template openings. The plaster mix is then screeded over the template shapes. Several layers are applied until the desired contour of the templates is attained. This concept is illustrated in Figure 5-7. Typical plaster molds are faced with metal deposition, polymer coating, or composite facing.

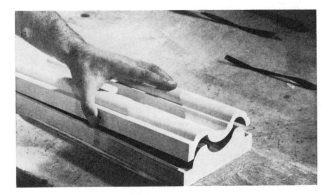

Figure 5-6. Accurate reproductions from master patterns are possible with low-cost plaster tooling. (U.S. Gypsum Company)

Male and female molds may be made simply by casting plaster over a model. The mix is poured into one corner of the supporting frame to reduce the possibility of entrapment of air bubbles or voids as the plaster flows across the model pattern. Vibration of the model pattern will also help to ensure reproduction of fine detail.

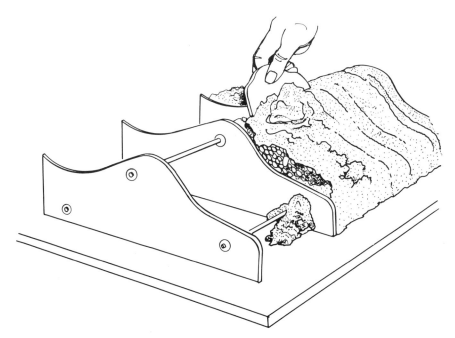

Figure 5-7. Plaster tooling using the loft template method.

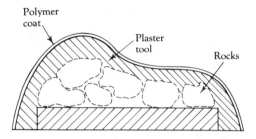

Figure 5-8. Plaster tooling over rocks offers an inexpensive tooling method.

Hand-shaped models may be produced by placing plaster over clay, wax, wood, or wire frame models. Even a pile of rocks may help to formulate a general shape. Once the plaster has been contoured and set, the rocks may be removed and replaced by reinforced plastics or metal framing. This concept is shown in Figure 5-8. Metal and plastic over concrete have been used successfully.

Closed, two-piece molds may be produced by casting a model pattern in a top and bottom half as illustrated in Figure 5-9. The two-piece mold may then be treated and prepared for casting, expanding, or other molding techniques.

Plasters are also used as hollow or solid mandrels in breakaway mold designs. The cores or special sections of the mold are broken or flushed away with high-pressure air and water. Specially fabricated hand lay-up ducting, pipe adapters, and other parts may be made by this technique.

Plaster tooling will continue to be used in tooling applications for production of small or large composite components or structures, especially since this tooling material lends itself so well to wet lay-up systems including bag methods. Plaster will remain an important master pattern material from which metal or polymer skin master molds are produced. Although they do not have desirable thermal conductivity, plaster master molds are excellent low-cost molds for thermoforming. Cooling coils are sometimes positioned and cast in the mold.

Polymer Tooling

Polymer tooling (plastics and elastomers) continues to find additional applications as a material for constructing molds. Applications are not limited to mold cavities. Polymer tooling is used to

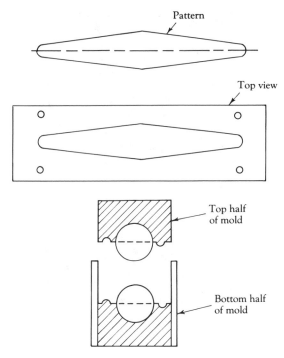

Pattern

Top view

Top half
of mold

Bottom half
of mold

Figure 5-9. Two-piece plaster tooling from master pattern.

make master patterns, transfer tools, cores, boxes, templates, draw
dies, jigs, fixtures, inspection tools, and prototypes.

Composite models are sometimes used to determine deformation
characteristics of new designs. These three-dimensional prototype
models must be made from materials corresponding to the final
product. This ensures that the prototype model simulates selected
conditions that the end product is expected to withstand. In Figure
5-10 this prototype chair is made of glass-reinforced polyester and
is used to determine points of maximum elongation. Necessary
changes to optimize design (e.g., addition of ribs, change in wall
thickness) are easily made in the prototype model.

Polymer tooling offers several advantages, including reduced development costs, chemical resistance, short manufacturing times,
low weight, and little or no machining.

A number of polymers are used with special emphasis placed on
specific processing techniques, stability, and dimensional tolerance
limits. Many must be reinforced or filled. Epoxy, phenolics, polyes-

Figure 5-10. Prototype model chair used for determination of points of maximum elongation. (BASF, Ludwigshafen)

ters, polyurethanes, polyamides, silicones, acrylics, and polyimide have been used. Aluminum and steel are common fillers that offer improved thermal conductivity, machinability, strength, and service temperature.

These molds are generally divided into two groups: those that are backed and those that are not. Many of the molds are made by casting or building up a composite layer over a model or pattern. (See Figure 5-11.) Cooling coils may be positioned and cast in place with a mixture of aluminum filler and epoxy. Pattern shapes may be made of wood, plaster, metal, glass, and honeycomb, solid, or cellular plastics. Epoxy-coated plaster tooling is used for casting, laminating, filament winding, thermoforming, and wet lay-up reinforcing techniques.

Large glass-reinforced epoxy molds of more than 100 ft in length have been used to produce composite hulls for ships. After the mold surface has been properly prepared with mold release and a gel coat, the composite structure is molded by lay-up and spray-up techniques. These large toolings are backed by a supporting frame and

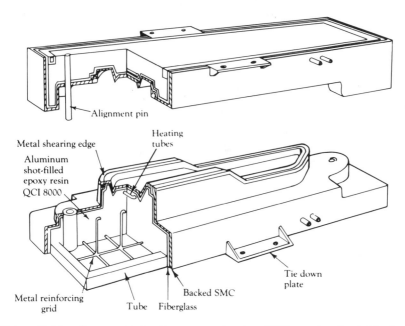

Figure 5-11. Composite tooling concept using SMC as a hard, stable, heat-resistant tooling product. (Quantum Composites, Inc.)

base. The tooling must be able to withstand prolonged exposure of curing temperatures in excess of 400°F (204°C).

Matched-die polymer tooling is used for forming mat or preforms, BMC, and SMC. Some laminating and sandwich construction is accomplished by this tooling process. A simple schematic of a matched-die tooling is shown in Figure 5-12.

Backed elastomers or plastics are used to produce numerous cast composite parts. Elastomers (mostly silicone and polyurethane) have high molding accuracy, flexibility, and excellent release action. Polymer materials may be cast, dip, or brush coated on wood, plaster, metal, or plastic patterns. A vacuum atmosphere is drawn on the elastomer mix to remove air bubbles. The cured layer is then removed and/or backed by various reinforcing media. Foams and honeycomb sheets are often used to provide strong, lightweight support for large tools. The prepared mold may then be used in some casting, lay-up, thermoforming, and expanding processes. The furniture industry uses silicone and polyurethane molds to duplicate wood grain designs. Many molds must remain flexible to provide faithful repro-

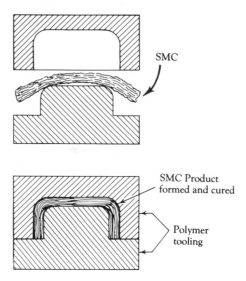

Figure 5-12. Polymer tooling may be used in matched-die molding of SMC.

duction of the master pattern and to provide easy removal from designs with undercuts.

Polymer tooling is used to provide accurate duplication of foundry patterns or core boxes. Composite patterns offer durable, tough, lightweight tooling for large designs. Polymers are also used to cast the plaster, wax, and ceramic cores for foundry use. Polymers have become a preferred material as master models, patterns, and duplication models for foundry and construction of some composite tooling.

Draw, stretch, hydroforming, and hammer dies are made of composite polymers. Most tools are backed with soft metals, high-density foam, or other composite materials. With the exception of hydroforming, forming tools are (normally) designed to resist deformation. Epoxy composites are used extensively as die faces. Other polymer composites, including phenolics, polyesters, and polyamides, are also used. Tough elastomers are selected for hydroforming and rubber-pad forming.

Wood Tooling

Wood tooling has been limited primarily to master patterns, fixtures, support frames, and short-run thermoforming and wet lay-

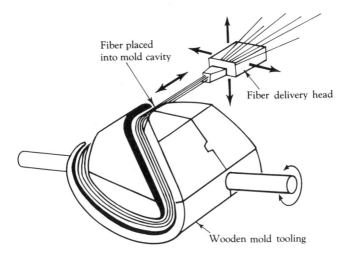

Fiber placed
into mold cavity

Fiber delivery head

Wooden mold tooling

Figure 5-13. Collapsible wooden tooling may be used in filament winding of this large composite frame structure.

up production. Polymer composite tooling has replaced many applications; for example, polymers are now used as mandrels and in low-pressure matched molding operations. Metal and polymer coatings are applied to improve the surface finish of the product and moldability. (See Figure 5-13.)

Miscellaneous Tooling

There are numerous and sometimes innovative applications of a variety of materials for tooling. For example, wax may be used in a lost wax technique. Once the composite is cured, the wax is heated (lost) from the mold surface or from hollow cores. Water-soluble salts are used in a similar fashion to plasters. Mandrels, bladders, or other mold shapes have been made of air-inflated elastomers. Elastomers are used in flexible-plunger and elastomeric press molding. A female cavity of metal and a flexible male plunger is used to shape composites parts; the basic concept is shown in Figure 5-14. In elastomeric press molding, an elastomer plug or sheet is used as a flexible half in a matched molding operation. A male or female metal mold may be used. A backing plate forces the composite and elastomeric material to conform to the mold contours. Titanium nitride, chromium carbides, and oxides have been used as coatings on molds, dies, and screws.

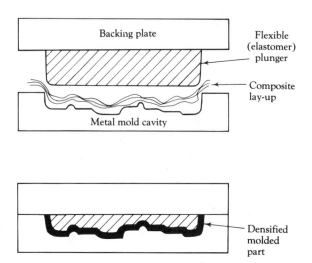

Figure 5-14. Flexible-plunger or elastomeric press molding.

Ceramic materials known as metalloids are extremely hard materials. They are made by powder metallurgy techniques and are used for some tooling when highly abrasive composite materials are processed. Alloys of vanadium, tantalum, or titanium carbide offer high wear resistance and a hardness of about 70 RC. Preshaping and curing dies for production line pultrusion operations sometimes use carbides or aluminum oxide dies or facings.

Solid carbon and silica ceramic materials including some glasses have been used in tooling. Carbon, which is easily machined, is used in mandrels or other cylindrical shapes. Ceramic materials have been used in thin shell-type tooling with metallic, polymer, and ceramic (glass) coatings. Glass may be blow molded into various mold shapes and has been used as mandrel shapes or female mold cavities for casting. After curing the glass may be broken and removed.

Chapter 6. What Are the Parameters for Designing Composite Products?

Introduction

It is beyond the scope of this book to summarize all the parameters for designing composite products. There are many texts and references devoted exclusively to composite design parameters. It is hoped that this chapter will serve as a fundamental guide and a useful starting point in understanding the complexity of composite designs. The reader should also note that basic design parameters vary greatly with the material and process selection.

Although there has been remarkable growth of composites in the past few decades, the industry is only in the developmental stages. Much of the research and development work has been accomplished for aerospace and military applications, although some civilian applications during the past two decades have resulted from the initial R & D work. An analogy of the composite industry may be made with the iron age. We are in a period of rapid technological growth and transition. Many improvements in the polymer matrices are made possible by alloying and the addition of numerous reinforcing agents. Pure iron is soft, easily bent, and has little strength, but with the addition of carbon and selected alloys, iron is transformed into a very durable, useful material. New designs, applications, and industries evolved from the development of composites. Like steel, composites are having and will continue to have a pervasive, dramatic impact on civilization. (See Product Applications.)

There are a number of trends that have changed the way we view composite materials. Most of today's applications require high and often sophisticated performance. Parts are being designed to use the merits of composite properties; they are no longer only substitutes for other materials. The major drawback to more rapid adoption of composites is a lack of design technology. The design considerations for composites are more complex than those of homopolymers or metals. The viscoelasticity of a polymer matrix cannot be compared with the elasticity of structural metals. An organic composite varies with time under load, rate of loading, small changes in temperature, matrix composition, material form, reinforcement configuration, and fabrication method. Isotropic materials have a well-defined elastic and plastic stress–strain behavior. Composites may be made isotropic, quasi-isotropic, or anisotropic depending on design requirements.

The principal advantages that may be gained from the use of organic matrix composites in design are the following:

1. Low energy costs per volume for manufacture and long-term energy savings from lighter components and structures are important economic considerations.
2. Parts may be designed to be anisotropic to exploit directional properties with specific strength and stiffness.
3. Labor costs are reduced by automation.
4. Fabrication processes allow the rapid manufacture of large, integrated, high-performance components.
5. Many designs provide excellent fatigue resistance.
6. Military and civilian applications have shown that many critical component applications can withstand a high degree of damage tolerance.
7. Corrosion resistance in a number of hostile environments is an asset for many applications.
8. Composites can be made with varying degrees of electrical and thermal conductivities.
9. The variety of materials, additives, blends, alloys, and processing techniques allows for greater flexibility in the design of most components or structures.
10. Many designs result in low scrap and postprocessing operations.

Basic Design Practice

The major emphasis in composite design and processing has been on automation, energy reduction, lowering scrap rate, increasing productivity, lowering labor costs, and improving reliability and processing techniques. Many of these concepts are being met by changes in process technology. It has been estimated that flexible manufacturing systems will account for more than half of all U.S. production by 1990. Increased utilization of computers will result in continued gains in productivity, product quality, and cost improvement. Programmable robots should be used to relieve monotony, gain repetitiveness, and protect workers from high noise and hazardous processes.

Great strides have been made with the introduction and use of CAD/CAM/CAE systems. The designer is able to use a computer in the design, engineering, and manufacturing of composite products. The entire production cycle may be studied using modeling and simulation. Some of the high-cost elements such as material handling, tooling, tool maintenance, raw material costs, and scrap losses may be evaluated and reviewed before production begins.

The data base may contain valuable information on the micromechanics and macromechanics of the matrix material selected. The detailed effects of temperature, stress, strain, time, and environmental exposure on composite stiffness, strength, and fracture toughness can be generated. The system may alert the user that a design is outside the parameters of the material or process selected. Interaction with the computer model and personnel will allow for accurate computations and design requirements. Productive models and computer-aided design and manufacture can play an active role in achieving polymer composite reliability.

The interrelationship of the complex design process is shown in Figure 6-1. This coherent design process may vary, with many functions overlapping. The order and sequence of events may vary at any phase in the development. Computer-integrated manufacturing (CIM) is a concept where all manufacturing processes are integrated, allowing designers, engineers, technicians, accountants, and others access to the same data base. The primary objective of CIM is to develop a cohesive digital data base that integrates the functions of manufacturing, design, and business operations. As a result, less human intervention is required on the shop floor. Part of this in-

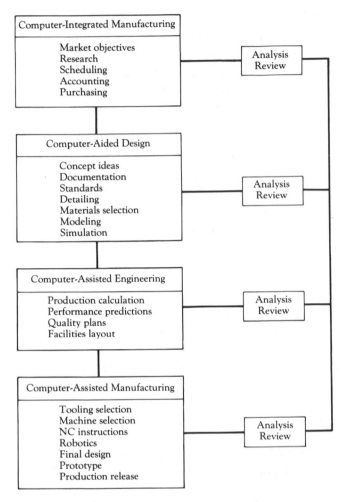

Figure 6-1. CIM, CAD, CAE, and CAM in the design production cycle.

tegration implies there would be benefits from flexible automation or manufacturing and savings in terms of reduced down time, labor costs, just-in-time (JIT) manufacturing or zero-inventory, quick changeover for batch manufacturing, and allowance for design changes. (See Tooling.)

Computer-aided design (CAD) is that part of the system that aids or assists in the creation, modification, and display of a design. It is used to produce three-dimensional designs and illustrations of the proposed part.

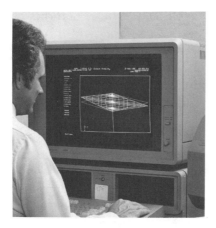

Figure 6-2. Designer uses finite-element analysis to test the behavior of a fiberglass highway sign under various stresses. (Owens–Corning Fiberglas Corporation)

Computer-aided engineering (CAE) is that part of the system that analyzes the design and calculates the performance predictions. Information from the CAD and CAM data is used to verify service life and safety design factors. (See Figure 6-2.)

Computer-aided manufacturing (CAM) is that part of the system that generates the manufacturing oriented data. CAM may involve production programming, robotic interfacing, quality control, and plant operations.

It is important to note that the design production cycle is an open-loop system with data being shared. A review and analysis connection is shown between each phase. This review procedure is conducted to confirm that all variables and constraints are considered in meeting the product objectives and requirements. The final design review is recommended prior to commitment to manufacture.

The design of composite parts involves three major considerations: (1) material, (2) production, and (3) design.

Material Considerations

In the past, the design was changed to meet the material limitations or properties. Today, materials must be selected with the right properties to meet design, economic, and service conditions. This is not always an easy task. It must be remembered that most matrix compounds are manufactured with a wide range of formulations and are available in a number of forms. Some are reinforced with particles or short fibers and sold in pellet form for injection molding, while others are available as preimpregnated (resin) fabrics, fibers, tapes, or compounds. Material form is usually associated with a specific production technique.

It cannot be assumed that the information obtained from data sheets or the manufacturer is adequate or predictive of matrix performance. Much of the data is based on laboratory-controlled evaluation. It is sometimes difficult to compare proprietary data from several different suppliers. Generally, values are derived from statistical models requiring the evaluation of numerous laboratory test coupons.

The following three requirements may be used as systematic screening methods of material selection for a specific design: (1) functional property factors, (2) processing parameters, and (3) economics. This does not imply that there is one best method to screen materials or that it is a step-by-step procedure. During analysis and selection, new information, technologies, or problems may require repetition of earlier steps.

Systematic methods are used with the aid of the computer. Failure analysis, cost-versus-property indices, and weighted property indices are familiar examples. In weighted property indices, each parameter is assigned a value depending on importance. The performance of materials may then be compared. Computer models can predict and anticipate most of the ways a material can fail. This method is called failure analysis. The complexity of composites and the combination of materials and processes compound the difficulty of failure analysis.

Functional Property Factors

One of the first requirements is to list (in quantitative terms) the **functional property factors** that the part is expected to tolerate. Examples of these properties are tensile strength, creep, thermal expansion, permeability, and impact strength. The following product property parameters are given as an example:

1. Relative density to be less than 2.6 g/cm^3.
2. Withstand repeated deflection of not more than 0.008 in. at 110°F.
3. Have a tensile strength of more than 20,000 psi.

4. Have a linear expansion of less than 10×10^{-6} in./in./°C.
5. Have a paintable, smooth surface.
6. Be able to withstand operating temperatures of more than 160°F.
7. Withstand repeated exposure in a hostile environment of petroleum fuel and saltwater.
8. Be able to withstand an impact of more than 1 ft-lb/in.
9. Withstand a service life of more than 10 years.
10. Meet UL standard 94 flammability of HB of not more than 1.5 in./min in 0.120–0.125 in. thick.

One question that should be asked is: "Do analogous applications exist?" A fully documented list of performance requirements may be available for a similar product.

The consumer is probably most aware of a product's appearance, utility, and reliability and is more likely to be interested in service life, wear resistance, ease of operation, ease of repair, and cost.

The composite designer must be concerned with structural and environmental questions at this phase. Everyone involved in the design operation must know where and how the matrix and finished composite component will be used.

The glass transition, melting, and crystallization of most matrices are reversible phenomena, while thermal degradation and cross-linking are not. There are limitations on the maximum service temperature to which organic matrix composites may be exposed. Figure 6-3 shows how various polymers perform on the basis of time and temperature. Materials from eight zones include the following:*

Zone 1

Acrylic
Cellulose acetate (CA)
Cellulose acetate butyrate (CAB)
Cellulose acetate propionate (CAP)
Cellulose nitrate (CN)
Cellulose propionate
Polyallomer
Polyethylene, low-density (LDPE)
Polystyrene (PS)
Polyvinyl acetate (PVAC)
Polyvinyl alcohol (PVAL)
Polyvinyl butyral (PVB)
Polyvinyl chloride (PVC)
Styrene–acrylonitrile (SAN)
Styrene–butadiene (SBR)
Urea–formaldehyde

*From D. V. Rosato, *Plastics World*, p. 30 (Mar. 1968).

Zone 2

Acetal
Acrylonitrile–butadiene–
  styrene (ABS)
Chlorinated polyether
Ethyl cellulose (EC)
Ethylene vinyl acetate
  copolymer (EVA)
Furan
Ionomer
Phenoxy
Polyamides
Polycarbonate (PC)
Polyethylene, high-density
  (HDPE)
Polyethylene, cross-linked
Polyethylene terephthalate
  (PETP)
Polypropylene (PP)
Polyvinylidene chloride
Urethane
Aromatic polyamines
Poly-para-xylylene
Polyaryl ether

Zone 3

Polymonochloro-
  trifluoroethylene (CTFE)
Vinylidene fluoride

Zone 4

Alkyd
Fluorinated ethylene propylene
  (FEP)
Melamine–formaldehyde
Phenol–furfural
Polyphenylene oxide (PPO)

Polysulfone
X-917 aromatic condensation
  polymer

Zone 5

Acrylic (thermoset)
Diallyl phthalate (DAP)
Epoxy
Phenol–formaldehyde
Polyester
Polytetrafluoroethylene (TFE)
Polybutadiene
Polybutadiene glycol
Polyphenylenesulfide
Polymethylenediphenyl oxide

Zone 6

Parylene
Polysulfone
Polybenzimidazole (PBI)
Polyphenylene
Silicone
Polybenzothiazole
Polyaryl sulfone

Zone 7

Polyamide–imide
Polyimide
S.A.P.
Polyphenylquinoxalines
Para-oxybenzoyl polymer
Polyquinoxaline
Polyimidazoquinazoline

Zone 8

Ladder polymers
Inorganic polymer network
Others to be developed

It is the matrix that degrades and results in composite failure.
Boron, graphite, carbon, and most ceramic and metallic fibers have

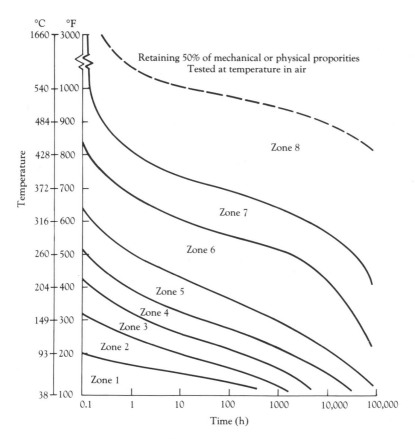

Figure 6-3. How plastics perform on the basis of temperature and time. [D.V. Rosato, *Plastics World*, p. 30 (Mar. 1968)]

high temperature resistance. Polyimide–boron fiber composites have high temperature resistance and high strength. Polyimide–graphite fiber composites can compete with metals in strength and achieve a significant weight saving at service temperatures up to 600°F (316°C). Epoxies in many epoxy–graphite fiber composites do not resist fire well. When exposed to excessive heat the epoxy matrix is damaged and releases the adhesion to the graphite fibers.

If the polymer matrix is to be exposed to intense heat or flame, only those materials that will not melt or seriously deteriorate should be evaluated. Boron powder may be added to the matrix to stabilize the char that forms in thermal oxidation. Other flame-resistant additives or an ablative matrix may be selected. The cryogenic temperatures of space or other environments are opposite thermal extremes. Radiation can cause cross-linking in some thermoplastic materials with a resulting change in properties. Weathering affects color, finish, transparency, and other properties.

If the component is for under-the-hood applications in automobiles, it must also be able to resist a chemical environment. Petroleum liquids, fumes, and salt solutions from the roadway may attack the matrix. Some polymers simply swell while others dissolve.

The laminar composites used for kitchen countertops and many furniture surfaces must also withstand chemical attack. Products requiring selective permeability and chemical resistance may select multilayered barrier films and packages.

Moisture may be absorbed into the matrix, resulting in deterioration and weakening of the reinforcement bond. Epoxies and polyimides are subject to moisture absorption which results in a loss of strength at elevated temperatures. Any exposure of reinforcement fibers greatly accelerates this potential problem. Allowance must be made for humidity and attraction of dust and dirt because electrical arcing over the surface may result. Surfaces or machined edges and holes may require a protective coating to prevent moisture infiltration or wicking. Carbon, graphite, and metal or metal-coated fillers and reinforcements may be beneficial in electromagnetic interference (EMI) shielding applications.

All the candidate materials for consideration can be subjected to the design parameters entered into the computer. This should yield a list of the possible matrix candidates which includes numerous matrix alloys and blends. Proprietary compounds or special mixtures of additives may be recommended. Do not be tempted to enter the cost of materials into your consideration at this phase. Materials with marginal performance properties and low cost should still be considered. Redesign may compensate or bring the part into line with specified requirements. Expensive materials can remain as possible candidates depending on the processing parameters and the economics of assembly and finishing.

At this point we should narrow the number of material choices to only a few possibilities (preferably less than six).

Processing Parameters

The next phase in the selection and systematic screening of a candidate matrix would include **processing parameters.** This task must consider many variables, including (1) the shape of the product, (2) tooling, (3) number to be produced (quantity), (4) rate of production, (5) capital investment for new equipment or technologies, (6) type and/or orientation of reinforcement, (7) required performance reliability and quality, (8) rheological (viscosity, flow) considerations of the matrix, and (9) economics.

There are a number of processing techniques used by industry to produce composite components and structures. Most of these processing techniques are listed in Table 6-1.

Table 6-1. Principle Composite Processing Techniques

Process	Remarks
Autoclave	Modification of vacuum and pressure bag; low production; low void content; dense parts; limited to autoclave size; wet
Blow molding	Mainly closed-mold process; multilayers; short fibers or particulates; high volume; small to medium-sized products
Casting (simple)	Open-mold process; low production; little control of reinforcement orientation; monomers or polymers
Compression molding	Closed-mold process; preforms available; some control on orientation; dense products
Expanding	Mainly closed-mold process; low to high production rates; limited control of reinforcement orientation; small to large products; monomers or polymers
Extrusion	Closed-mold process; continuous lengths; multiple layers; continuous long fibers possible; preforms possible; some control on orientation
Filament winding	Mainly open-mold process; low production; control of orientation; wet
Hand lay-up	Open-mold process; low production; control of reinforcement orientation; large parts; wet
Injection molding	Closed-mold process; short fibers or particulates; little control on reinforcement orientation; small to medium-sized products

Continued

Table 6-1. Continued

Process	Remarks
Laminating (continuous)	Mainly closed-mold process; medium to high production; control of reinforcement orientation; small to large products; continuous; monomers or polymers
Matched-die molding	Closed-mold process; low to high production; some control of orientation; preforms; medium-sized products; monomers or polymers
Mechanical forming	Mainly closed-mold process; medium to high production; preforms; some control of reinforcement orientation
Pressure-bag molding	Open-mold process; low production; control of reinforcement orientation; preforms; wet
Pulforming	Mainly closed-mold process; continuous; some control of reinforcement orientation; preforms; monomers or polymers
Pultrusion	Mainly closed-mold process; continuous; some control of orientation; preforms; monomers or polymers
Reaction injection molding	Closed-mold process; small to medium sized products; medium production; monomers or polymers
Rotational casting (centrifugal)	Open-mold process; low production; little control of reinforcement orientation; small to large products; powders and wet
Spray-up	Open-mold process; low production; little control of reinforcement orientation; preforms; wet
Thermoforming	Mainly open-mold process; preforms; medium to high production; little control of reinforcement orientation; mostly short fibers or particulates
Transfer molding	Closed-mold process; high production; dense parts; little control of reinforcement orientation; small to medium products; short fibers or particulates; monomers or polymers
Vacuum-bag molding	Open-mold process; low production; control of reinforcement orientation; preforms; wet

Shape of Product

The physical **shape of the product** may narrow the matrix choice to only a few. For example, it may not be practical to select polyolefin composites for a marine hull, but they may be considered for septic tanks. Epoxies and polyesters are considered the workhorses of the composites industry. They are used in the production of boat hulls, aircraft sections, helicopter blades, drive trains, and other structural parts. Some composite materials are more easily produced into small, intricate, and complex shapes than others.

In conjunction with shape, tooling must be considered. It must be ascertained if it is practical to shape the matrix on conventional tooling. Polyamide with short fibers or particulates is successfully molded into small parts in automobiles, power tool casings, and components. Heavily filled and reinforced composite materials may require special precautions and preparations in tooling.

Quantity

If a large **quantity** of parts are to be produced, some materials may not be practical. Some are not easily processed, or a great deal of reinforcement may be needed to obtain the required mechanical properties. These additives may result in molding problems. In addition, the increased viscosity may require a different material or processing technique to be selected.

Rate of Production

The **rate of production** may require that only fully polymerized thermoplastic materials be used; for example, perhaps only those materials that lend themselves to full automation can be selected. This would eliminate most wet-type processes. Monomeric materials are often used where larger shapes or lower production volume is needed.

Capital Investment

Shape, tooling, quantity, and production rate criteria may determine the selection of a material that requires additional capital investment for **new equipment,** such as jigs, fixtures, or curing racks. The physical space of producing large composite structures in a controlled atmosphere such as an autoclave may result in consideration of different materials and/or processes. Many molders specialize in one or two molding techniques which limits the potential materials selection.

Reinforcements

The **type** and **orientation of reinforcement** may not be available in the desired matrix formulation. Fiber- and particulate-reinforced materials are available in a number of forms, from pellets to preforms. If the product requires specified fiber orientation, a monomeric preform or impregnating system may be needed. The use of LRIM over RIM processes permits heavy reinforcements and mass production. The surface condition of the reinforcement and matrix adhesion can be critical to the production of composites of high quality. Carbon-reinforced composites are brittle and have no yield behavior. In addition, the matrix expands more than other materials; consequently, there may be thermally induced stresses. Compatibility must be considered any time several different materials are combined. Thermal expansion and galvanic corrosion and electrolysis may present a problem with metallic reinforcements or assemblies. (See Reinforcements.)

A matrix candidate with minimal performance characteristics may not be the best choice if **reliability** and high **quality** are important. Reliability of a proposed composite material is difficult to measure because it is dependent not only on material properties but on design and processing. The moisture sensitivity of some E-glass composites may be overcome by the proper choice of resin systems and ratio of curing agent. Postprocessing surface treatments may be required. Heavily reinforced or cellular composites sometimes result in characteristic flow marks, swirls, and weld lines. Matrix, design, and processing greatly influence the appearance of the product. Moisture and corrosion of the composites can result in notable decreases in the strength and thermal resistance of the matrix. Moisture acts as a plasticizer in epoxy and polyimide polymers.

Rheological

Closely associated with the material form and production process are the rheological considerations of the matrix. Reinforced fluoroplastics may have all the desired performance properties but some are not melt processible. Heavily filled, viscous materials may not be easily forced into the mold cavity, resulting in poor surface quality. Mold temperatures, melt temperature, and flow speed (injection) all have a significant effect on the gloss of molded parts. Rapid filling of a mold cavity also minimizes fiber orientation and enhances weld-line integrity. Thixotropic additives may greatly aid processing because the matrix is gel-like at rest but fluid when agitated. The properties of a matrix affect the wetting, reaction, compatibility, and stress transfer to the reinforcements.

Economics

The final phase in material selection is to consider the economics. For many, cost may be the most important single factor in selecting suitable materials for the composite product. Some matrix materials or reinforcements with the most desirable properties may be too expensive to market. For composite products to compete with other materials in a variety of applications, companies must be able to sell consumer products at a profit. Better education, planning, design, and the use of newer technological operations, including automatic fabrication of composite parts, should improve productivity, quality, reliability, and profitability. Many wet or open molding operations are labor intensive and cannot hope to compete with companies that have increased production of composite parts in automated facilities. The CAD/CAM/CAE/CIM systems are essential for cost-effective, large-scale production. In addition to the development and production of superior composite parts, these systems may reduce materials handling and inventory and maximize utilization of equipment and personnel.

There are a number of methods to enhance the economic attractiveness of the matrix. It is possible to change or optimize the micromechanics of the matrix material. Additives may increase the desired property. Remember that a change in one property may result in a change in several others; for example, the addition of glass fillers may increase the chemical and thermal resistances but reduce toughness and rheological properties. Diluting expensive matrices with fillers or other additives may provide a cost advantage and improve or maintain performance properties.

Cost is often based on the production method and the number of items to be produced. For example, a one-piece seamless gasoline tank may be rotationally cast or blow molded. The latter process uses more costly equipment but can produce the tanks more quickly, thus reducing costs. A large storage tank may be produced at less cost by rotational casting than by blow molding. To amortize the tooling cost, volume sales are needed. For example, if the tooling costs were $50,000 and only 10,000 parts were made, the tooling cost would be $5.00 per part. If 1,000,000 parts were made the tooling costs would be $0.05 per part. Some processing operations may require a special atmosphere or protection for personnel. One material may be more costly because it is more difficult to machine, fabricate, or finish. A comparison of processing and economic factors is shown in Table 6-2. It should be obvious that equipment and tooling costs will vary depending on part size, performance needs, and complexity of design.

Composite materials are expensive when compared with other materials on a per volume or weight (mass) basis. Composite parts may cost from $2.00 to $200 per kilogram. On a cost per kilogram

Table 6-2. Economic Factors Associated with Different Processes

Production Method	Economic Minimum	Production Rate	Equipment Cost	Tooling Cost
Autoclave	100–1,000	Low	High	Low
Bag molding	100–1,000	Low	Low	Low
Blow molding	1,000–10,000	High	Low	Low
Casting processes	100–1,000	Low–high	Low	Low
Compression molding	1,000–10,000	High	Low–high	Low–high
ERM	100–1,000	High	Low–high	Low
Expanding processes	1,000–10,000	High	Low–high	Low–high
Extrusion (meters)	1,000–10,000	High	High	High
Filament winding	100–1,000	Low–high	Low–high	Low–high
Injection molding	10,000–100,000	High	High	High
Laminating	1,000–10,000	Low–high	Low–high	High
Lay-up	100–1,000	Low	Low	Low
Matched die	1,000–10,000	High	High	High
Press molding	100–1,000	High	Low–high	Low
Pultrusion (meters)	1,000–10,000	High	High	High
RIM, LRIM, RRIM	1,000–10,000	High	High	Low–high
Rotational casting	100–1,000	Low–high	Low–high	Low
RTM	1,000–10,000	High	Low–high	Low–high
Thermoforming	100–1,000	High	Low	Low
Transfer molding	1,000–10,000	High	High	High

basis, the matrix may cost ten times more than steel; yet on a volume basis, some are lower in cost than steel.

Apparent density and bulk factors are sometimes used to compare the costs of different matrices.

Apparent density (bulk density) is the weight (mass) per unit volume:

$$\text{apparent density} = \frac{W}{V}$$

where

V = volume (in cm³) occupied by the material in the graduated cylinder, HA

H = height (in cm) of the material in the cylinder

A = cross-sectional area (in cm²) of the measuring cylinder

W = mass (in grams) of the material in the cylinder

Bulk factor is the ratio of the volume of loose molding powder to the volume of the same mass of matrix after molding.

$$\text{bulk factor} = \frac{D_1}{D_2}$$

where

D_1 = average density of the molded or formed specimen

D_2 = average apparent density of the matrix material prior to forming

Composite parts may provide long-term benefits. Cost-effectiveness calculations must consider that many composite designs and processing techniques will result in a net savings by reducing the number of parts and the need to fabricate or assemble many components. One-piece hulls, fuselage, or floor pan for an automobile may greatly reduce multiple tooling and assembly of components. Reduced weight can more than offset the higher-cost material by decreasing fuel costs. It has been estimated that for every 100 lb removed from a 2800 lb automobile, there will be a fuel savings of 0.3 mi/gal (or mpg). These composite parts may also reduce corrosion, dampen sound vibrations, reduce thermal transmission, and improve fatigue properties. For example, a composite transportation vehicle may have an increased service life.

Since composites are mostly petroleum derived, it is apparent that competition for raw materials will continue. Any material selection must consider the availability of the material resources. For example, matrix and reinforcing materials may become scarce if one company or country withholds the materials from the market.

A principal advantage of using organic matrix composites in many transportation designs is the lower energy cost. This represents lower energy cost for a given volume of fabricated composite and lower energy cost for fuel during the lifetime of the composite part. For example, a composite automobile body may reduce the total energy cost over a steel body by more than 40%. The energy requirement of selected polymers and metals is shown in Figure 6-4.

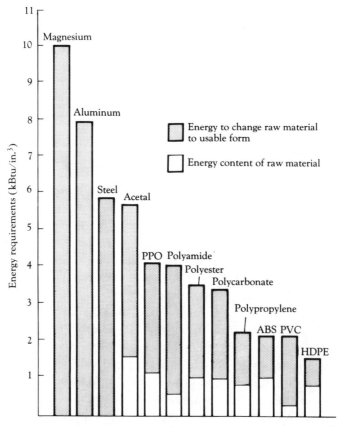

Figure 6-4. Energy requirements of selected polymers and metals.

An increasingly important factor in materials selection is disposability and recyclability because of concern about the environment. Many composites do not degrade easily when compared with metals. Only a small portion of the materials used by the composite industry is economically recyclable. Homopolymers are more easily reused than heavily reinforced and filled polymers. In some operations such as injection molding and extrusion, a percentage of regrind may be used as filler. Although it is difficult to visualize how a composite boat hull or automobile fender could be recycled into useful products, this does not mean that research could not provide an economical method and a useful product. One obvious method of disposal would be to incinerate these organic materials and capture the energy that they contain. High-technology incineration of municipal waste, including composite parts, could provide a safe and relatively pollution-free energy source for generating electrical power.

Standards have a direct and indirect impact on economic considerations. (See Standards.)

Product Considerations

The part shape, size, matrix formulation, and matrix form often limit the means of production to one or two possibilities. Although selecting the method of production appears relatively straightforward, a number of parameters must be considered. The feasibility of making the tooling, the capacity of the molding equipment, and the material exhibit a close relationship. Details such as forming pressure, temperature, surface quality, postcuring cycles, and production rates all come together in selecting the processing method.

As a rule, short-fiber- and particulate-reinforced materials are molded by high-production methods such as injection molding. These molded parts are not as strong as parts molded by processes that allow controlled filament orientation. Continuous fiber methods produce a highly aligned fiber arrangement yielding high-quality, nearly ideal composites. Pultrusion, filament winding, and laminating are typical processing methods.

Large composite structures such as pressure vessels, helicopter blades, fuselage structures, spacecraft parts, and storage tanks may require considerable investment in equipment and tooling. Large (> 2000 gal) multilayered, isotropic, reinforced storage tanks have been blow molded and rotationally cast.

Any postfabrication or finishing operations must also be considered as part of the production operation. If the composite product requires surface treatments such as painting, shielding, or other coating operations, alternate production techniques may be required. Trimming, cutting, boring holes, or other fabricating or assembly techniques may slow production, lower performance properties, and increase costs.

Design Considerations

Designing composite components or structures is an extremely complex activity. Many are designed to be anisotropic to fully benefit from the directional properties of the reinforcing additives. Composites with short fibers or particulates may be relatively isotropic or designed for anisotropic orientation of properties. Some crystalline polymers are processed with a directional molecular orientation.

In planning the preliminary ideas for a composite design there are a number of features to keep in mind: (1) overall design parameters that have been produced from design studies, (2) overall design conditions that the part must meet, (3) tooling parameters, and (4) design analysis.

The advantages of composite materials are becoming more familiar to a wider range of designers and consumers. Composites have allowed designers to develop components and structures not possible with isotropic materials. One of the earliest steps in product design is to review the data and design studies that have been conducted. These may provide valuable information and help in the preliminary design of a new product.

Many problems may be overcome if the designer is well informed. Composites simply do not perform like isotropic and homogeneous materials which have well-defined elastic and stress–strain properties. (See Figure 6-5.) The characteristic properties of a composite are derived from its constituents, from its processing, and from its microstructure. Experience plays an important part in the understanding of composite materials and processing.

In considering the overall design conditions, the intended application or function, environment, reliability requirements, and specifications must be reviewed. It is helpful to list the anticipated conditions, use, and performance requirements that the part is likely to be expected to withstand. During the preliminary design review,

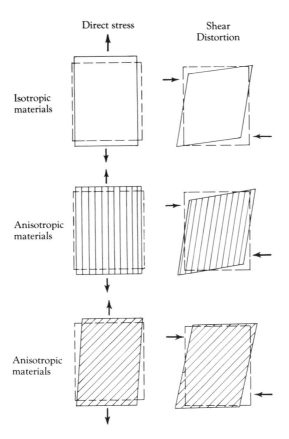

Figure 6-5. Tensile stress causes lengthening and contraction. Shear stress causes angular distortion.

it may be necessary to consider the performance benefits of weight, cost, and property trade-offs. Deliberate strength anisotropy may result in increasing glass content or wall thickness. This generally results in weakening at right angles to the fibers. The reduction of weight (mass) may be desirable but requires more expensive materials and processing. Care must be taken during this stage that generalized views and data do not mislead the designer. The complexity of designs and sensitivity to material prices and innovative production techniques may alter many decisions.

Part of the preliminary design must include tooling parameters. Next to materials selection, tooling and processing have a marked effect on the properties and quality of composites. Properly designed tooling may minimize stress concentrations and maximize desirable properties. Tolerance specifications may demand that closed molding methods be used. It may be necessary to develop the composite design for an appropriate method of production. Since in industry a major emphasis is placed on design to cost, the composite must be designed for rapid production technologies.

For composite materials to gain increased use, they must be cost effective with other material choices. There have been significant savings in assembly and tooling costs in the automobile industry by new designs and integration of parts. Integrated composite structures are presently made into bath fixtures, aerospace panels, and aircraft and automotive parts.

Design Analysis

Design analysis is most complicated for polymeric composites. There is a large data base of properties needed for the design and analysis of the part over a variety of conditions. The amount of calculations and data require that a computer be used. Finite-element analysis is a method to analyze the stress response in structural elements. The part to be studied is divided into elements which are jointed by node points. This network of nodes is used to show how stresses are transmitted. By using the computer model it is possible to show the stress response of the part with a specific geometry. CAD systems greatly reduce the model construction time required. Finite-element analysis is only one step in the total product development process.

Safety Factor

A designer must produce a plan that will satisfy cost, functional, and reliability requirements. It is more important to be able to predict property or part failure. A safety factor (sometimes called design factor) is defined as the ratio of the ultimate strength of the material to the allowable working stress:

$$SF = \frac{\text{ultimate strength}}{\text{allowable working stress}}$$

Allowable safety factors depend on a number of variables, many of which are specified by codes or recognized authorities. The safety factor value should be based on (1) accurate, reliable load estimates;

(2) analysis and stress determination, (3) expected adverse environmental conditions, (4) quality of the processing technique to produce reliable parts, (5) the nature and inhomogeneity of the loads, and (6) criticality of application.

High safety factors have been used in designing polymeric composites for years. This overdesign has been a result of inexperience. While metals have a well-established and demonstrated performance level, composites lack this homogeneity. All values of composites are a function of many variables and testing methods.

The safety factor for many composites is 4.0. For critical structural composites, this factor may be as high as 10.0. Obviously, a composite helicopter rotor blade is a more critical application than the drive link in an automobile.

With accurate, reliable data some designs are using a safety factor of 1.5. This means that there must be a high degree of confidence in the method of fabrication, quality of materials, available methods of testing, and knowledge of loads. Weight savings is a common goal in most composite designs. If the safety factor and weight are to remain low, proper design and processing are key elements. Many of the aerospace processing techniques are extremely labor intensive to assure that design and processing defects are minimal.

The ultimate test is service. Computer modeling and other aids are simply not sufficient for some designs. It may be necessary to test functional models and prototypes to optimize the design and determine performance. Even if the safety factor is not a major concern, the quality of parts may warrant a prototype. A prototype mold is invaluable in providing answers to questions on the molding needs of the part. This information can then be incorporated into the production mold.

Basic Design Practices

There are numerous sources of information concerning basic design practices: Volumes have been written on the subject. Because of the present space limitations, only two broad categories of guidelines are discussed: (1) product design and (2) mold or tooling design. This does not imply that they are to be considered as separate concepts. The very fact that a part is to be molded, not cast or machined, should illustrate the interdependence of design considerations.

General product and mold design guidelines must be further divided into the three major classifications of composites: (1) fibrous, (2) laminar, and (3) particulate.

Fibrous Composite Design Guidelines

As a rule, we know that components benefit in several performance areas from fibrous reinforcements. This can be seen in selected properties of unreinforced versus (30%) glass-reinforced polymers in Table 6-3. Thermosetting compounds have similar improvements.

The following generalized guidelines may be applied to most thermosetting and thermoplastic materials molded by injection, compression, and transfer molding techniques. Some may also apply to other processing techniques.

Standard stress and deflection formulas may be calculated under various loading conditions at room temperature. Figure 6-6 shows the beam dimensions and calculation of the moment of inertia (I).

W = load (lb)
L = length of beam between supports (in.)

Table 6-3. Selected Properties of Reinforced
Versus Unreinforced Thermoplastic Polymers[a]

Polymer	Mold Shrinkage (in./in.)	Tensile Strength (10^3 psi)	Thermal Expansion (10^{-5} in./in. $-$ °F)	Deflection Temperature at 264 psi (°F)
Acetal	0.003 (0.020)	19.5 (8.8)	2.2 (4.5)	325 (230)
Polyamide 6/6	0.004 (0.016)	22.0 (11.8)	1.8 (4.5)	485 (170)
Polycarbonate	0.001 (0.006)	18.5 (9.0)	1.3 (3.7)	300 (265)
Polyester (PBT)	0.003 (0.020)	19.5 (8.5)	1.2 (5.3)	430 (130)
Polyetherimide	0/002 (0.006)	28.5 (15.2)	1.1 (3.1)	420 (392)
Polyetheretherketone	0.003 (0.011)	25.0 (14.5)	1.8 (5.0)	600 (360)
Polyether sulfone	0.003 (0.007)	19.0 (12.0)	1.8 (3.1)	415 (400)

[a]Values in parentheses are for unreinforced thermoplastic polymers.

c = distance from the outermost point in tension to the neutral axis (in.)

b = beam width (in.)

d = beam height (in.)

E = modulus of material example

S_{max} = maximum stress

Y_{max} = maximum deflection

C = cyclic stress of material example

I = moment of inertia (in.^{4})

Z = section modulus

M = load × distance to support (in.-lb)

Example: Moment of Inertia

$$I = \frac{bd^3}{12} = \frac{(0.25 \text{ in.})(0.50 \text{ in.})^3}{12} = 0.0026 \text{ in.}^4$$

Example: Steady Load (Creep)

$$E = 17,000 \text{ lb/in.}^2$$

$$W_{max} = \frac{4S_{max}I}{Lc} = \frac{(4)(17,000)(0.026)}{(3.0)(0.25)}$$

$$= 236 \text{ lb}$$

Example: Cyclic Load (10^7 Cycles)

$$C = 4550 \text{ lb/in.}^2$$

$$W_{max} = \frac{4S_{max}I}{Lc} = \frac{(4)(4550)(0.0026)}{(3.0)(0.25)}$$

$$= 63 \text{ lb}$$

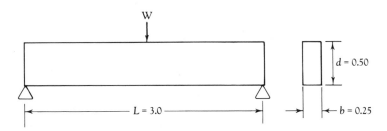

Figure 6-6. Beam used in examples.

Example: Short-Term Loading

$$L/2 \text{ and } M = \frac{WL}{4}$$

$$S_{max} = \frac{WLc}{4I}$$

$$E = 17{,}800 \text{ lb/in.}^2$$

$$W_{max} = \frac{4S_{max}I}{Lc} = \frac{(4)(17\ 800 \text{ lb/in.}^2)(0.0026 \text{ in.}^4)}{(3.0 \text{ in.})(0.25 \text{ in.})}$$

Short-term working stress = 247 lb

$$Y_{max} = \frac{WL^3}{48EI} \text{ at } L/2$$

$$E = 15.6 \times 10^5 \text{ lb/in.}^2$$

$$= \frac{(247 \text{ lb})(3.0 \text{ in.})^3}{(48)(15.6 \times 10^5 \text{ lb/in.}^2)(0.0026 \text{ in.}^4)}$$

$$= 0.034 \text{ in.}$$

Additional examples of load consideration are shown in Figure 6-7, 6-8, and 6-9.

For a given composite, creep strain is directly related to the applied load. Creep resistance is particularly important for extended service. In a thermoplastic matrix composite, creep rate is inversely related to the amount of fiber it contains although not proportionally to stress. Flexural creep is shown in Figure 6-10.

Integrated part designs are most efficient and generally reduce overall part cost. If the designer can combine several functions and components into a single molded part, such as grills, ribs, or brackets, assembly and additional tooling costs can be reduced.

Design efficiency may require a decrease or an increase in wall section. Composite designs may have reduced wall thickness up to 50% over unreinforced materials. However, reduction in wall thickness may not always be the proper choice for decreasing costs. The use of ribs, contours, corrugations, and other geometric factors may be a better alternative. When a composite design requires a varying wall thickness, gradual transition is recommended to eliminate distortion and reduce internal stresses. Part geometry is directly related

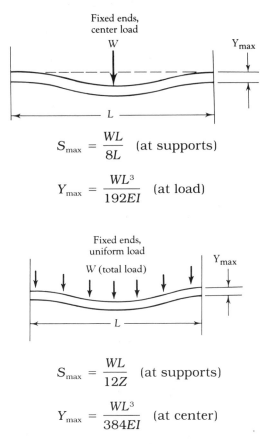

$$S_{max} = \frac{WL}{8L} \quad \text{(at supports)}$$

$$Y_{max} = \frac{WL^3}{192EI} \quad \text{(at load)}$$

$$S_{max} = \frac{WL}{12Z} \quad \text{(at supports)}$$

$$Y_{max} = \frac{WL^3}{384EI} \quad \text{(at center)}$$

Figure 6-7. Examples of load considerations.

to how the matrix will fill the mold. This affects appearance, cycle time, flatness, dimensional stability, and other performance properties of the part. Wall transition is shown in Figure 6-11. Fiber-reinforced polymers shrink more along the axis transverse to flow than along the axis of material flow. Because some fibers are broken during the molding process, careful mold design is important to ensure some degree of control over fiber length and orientation.

To facilitate removal of the part from the tooling, a small draft angle may be required. Draft angles vary from 0.5° to 3° depending on complexity of design, depth of draw, and texture of the mold surface. This is especially true with cores. With textured designs, the draft angles should be at least 1° per side (inside and outside)

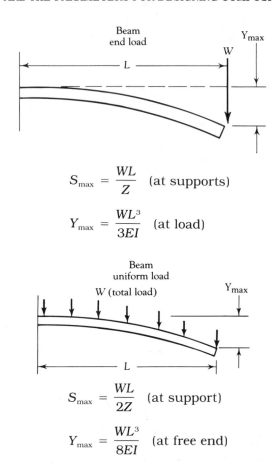

$$S_{max} = \frac{WL}{Z} \quad \text{(at supports)}$$

$$Y_{max} = \frac{WL^3}{3EI} \quad \text{(at load)}$$

$$S_{max} = \frac{WL}{2Z} \quad \text{(at support)}$$

$$Y_{max} = \frac{WL^3}{8EI} \quad \text{(at free end)}$$

Figure 6-8. Examples of load considerations.

for every 0.001 in. of depth. Typical shrinkage of fiber-reinforced polymers is about one-third to one-half that of nonreinforced polymers.

Warpage is somewhat proportional to the amount of shrinkage of the matrix. Residual stresses are developed as a result of forcing the material to conform to a mold shape (molecular and fibrous orientation). During cooling or curing, matrix shrinkage also locks in stresses. These stresses make it difficult to produce an absolutely flat surface in molded parts. Straight side walls or surfaces may be designed with a slight dome of 0.002 in. or more to improve the surface appearance, strengthen the part, and resist warpage.

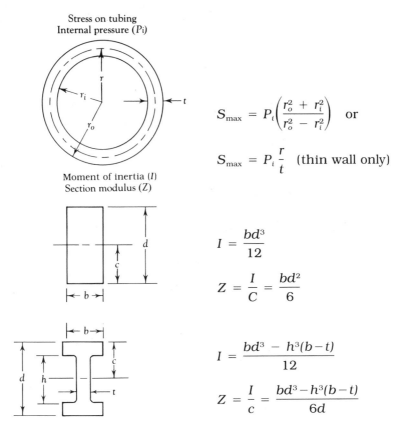

$$S_{max} = P_t \left(\frac{r_o^2 + r_i^2}{r_o^2 - r_i^2} \right) \quad \text{or}$$

$$S_{max} = P_t \frac{r}{t} \quad \text{(thin wall only)}$$

$$I = \frac{bd^3}{12}$$

$$Z = \frac{I}{C} = \frac{bd^2}{6}$$

$$I = \frac{bd^3 - h^3(b-t)}{12}$$

$$Z = \frac{I}{c} = \frac{bd^3 - h^3(b-t)}{6d}$$

Figure 6-9. Examples of load considerations.

Coring is an effective way to reduce heavy sections, when the heaviness is not needed for strength. Every effort should be made to keep all coring in the direction of pull or parallel to movement of the mold when it opens. Blind cores should be avoided. As a rule, cores less than 3/16 in. diameter should be no greater than twice the diameter. Cored-through holes should not exceed six times the diameter. Coring recommendations are shown in Figure 6-12.

Ribs are used to reduce wall thickness yet support the desired loads of the part. Thick, heavy ribs may cause vacuum bubbles or sink marks at the intersection of surfaces, therefore thin ribs are preferred. In general, rib size should have a width at the base equal to one-half the thickness of the adjacent wall. They should be no

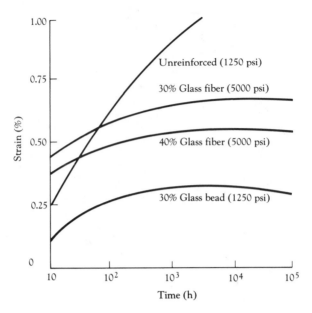

Figure 6-10. Creep (73°F) of reinforced polyamide 6/6.

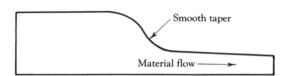

Figure 6-11. Gradual blending between different wall thicknesses.

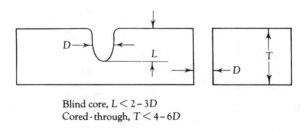

Figure 6-12. General coring recommendations for fibrous-reinforced parts.

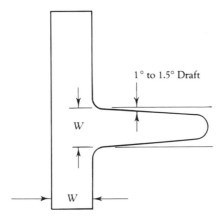

1° to 1.5° Draft

W

W

Figure 6-13. General recommendations for rib size and design.

higher than three times the wall thickness. A taper should be used. Some sink marks may be eliminated by locating a sprue or gate close to the ribbed section. Rib design recommendations are shown in Figure 6-13. General practice is to gate into the thickest section of the part to minimize sinks or voids. Fibers may be damaged if gated against a cavity wall or core pin. Round or rectangular gate size should be equal to the full width of the cavity wall. Full round runners of more than ¼ in. in diameter or the wall thickness of the part are recommended.

Flash forming around wedges, knockout pins, plugs, or name plates should be easy to clean.

Bosses are commonly used to facilitate mechanical assembly and add support. Generally, the outside boss diameter should be equal to twice the inside diameter of the hole. They should be no higher than twice their diameter. Several boss designs are illustrated in Figure 6-14.

Undercuts require collapsing or removable cores and should be avoided to minimize tooling costs.

Sharp corners, fillets, and radii should also be avoided. Sharp corners result in stress concentrations, while liberal fillets and radii on internal corners reduce stress concentrations. A radius equal to half the adjacent wall thickness is the recommended minimum.

Both internal and external threads may be molded into fibrous composite parts. Metallic inserts molded into the part have greater strength. Generally, the ratio of wall thickness around the insert to

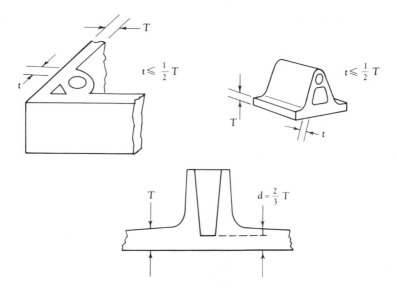

Figure 6-14. Selected boss designs.

the outer diameter of the insert should be slightly greater than one. This will allow sufficient material for strength. Any time two different materials are used (metal/polymer), allowance for differential expansion must be considered.

There are a number of design variables when holes are required in the part. Fibers tend to orient with a resulting weld line around holes. This presents a potentially weak point. Proper gating and rapid filling are important to avoid this weld line.

Because of the high viscosity of most fibrous matrices, the mold cavity should be vented wherever air may become entrapped. Vents of 0.004 in. long × 0.200 in. wide are sometimes used.

Louvers or grillwork should be oriented in the direction of material flow. If this is not possible, a runner should be used across the middle of the grillwork to allow easy filling of the grill from the center toward the sides.

Assembly methods vary greatly. There are a variety of mechanical joining techniques available for fibrous-reinforced composites, varying from snap fits to mechanical fasteners such as screws and rivets.

Assembly methods for ultrasonic, solvent, and adhesive bonding are well established. Remember, assembly is associated with stress, both chemical and mechanical. Care should be taken to minimize assembly stresses. Washers are used under bolt and screw heads to help distribute the compression forces caused by torque assembly. Self-tapping screws may be used with most thermoplastic and thermoset materials.

A snap-fit assembly, an economical and simple method of joining many composite parts, is generally used less than ten times and little or no stress is left on the flexing finger after being snapped into place. (See Figure 6-15.) Graphite fiber composites are not suitable for snap-fit assembly. No finger beam should be expected to

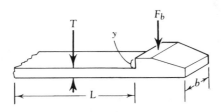

$$F_b = \frac{yEbT^3}{4L^3}$$

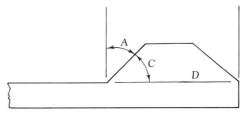

where E = flexural modulus

F_b = bending force

C,D = incident angles

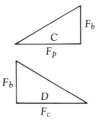

Pullout force (F_p):

$$F_p = \frac{F_b}{\tan(90° - C)}$$

Insertion force (F_i):

$$F_i = \frac{F_b}{\tan(90° - D)}$$

Figure 6-15. Snap-fit assembly forces.

exceed the recommended dynamic strain limit. Figure 6-16 shows how to calculate straight- and tapered-finger beam strain. The difference between a snap and press fit is the undercut.

In blow molding and thermoforming processes, the material is stretched at temperatures below the melt temperature. It should be apparent that these processes result in stretch orientation of the material and produce frozen-in stresses. The susceptibility of the part to damage correlates to the extent of stretching and orientation. Designs for these processes should attempt to minimize the amount of stretching.

Part design for **structural foamed** composites are similar to injection molding. The cellular core and solid skin of structural foam parts are made by both low- and high-pressure processes. (See Expansion Processes.) In low-pressure methods (< 1000 psi), there is very little molded-in stress. A higher skin to cellular core ratio generally results in increased strength properties. Because of the cellular core, bosses should have interconnecting ribs or gussets to distribute loads. Ribs should be much heavier than those in solid configurations. In general, ribs should have a width at the base equal to the thickness of the adjacent wall. To avoid sink marks and warpage problems, a uniform wall thickness should be maintained when possible.

Composite parts may be **extruded.** The design problems are similar to pultrusion processes. Both thermosetting and thermoplastic matrices can have numerous reinforcements, including continuous filaments and fibers. The screw action results in some fiber damage if plasticated with the matrix. Pull-back and shrinkage must be compensated for if dimensional tolerances and profile shapes are to be of high quality. Longer forming dies and slower output rates may help improve this problem. Extruded parts have molecular and reinforcement orientations produced by the drawdown process, resulting in frozen-in stresses.

Casting processes (e.g., simple, rotational, slush) are used to produce true composites using little or no filling pressure (atmospheric pressure). Relatively stress-free parts are produced with little or no waste (scrap). Bubbles or porous sections are a frequent problem with casting operations, and powders or viscous materials may have difficulty in filling all parts of complicated molds. Vacuum chambers are sometimes used to remove unwanted bubbles from liquid polymer mixtures prior to curing. The particle size and poly-

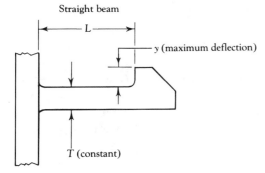

Straight beam

For rectangular cross sections:

$$e_D \text{ (dynamic strain)} = \frac{3yT}{2L^2}$$

For any cross section:

$$e_D = \frac{3yc}{L^2}$$

where c = distance from neutral axis to extreme fiber.

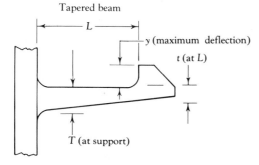

Tapered beam

For rectangular cross section:

$$e_D = \frac{3yT}{2L^2 k_p}$$

For any cross section:

$$e_D = \frac{3yc}{L^2 k_p}$$

where c = distance from neutral axis to extreme fiber at support

k_p = stress-concentration factor

Figure 6-16. Straight- and tapered-finger beam calculations.

273

mer viscosity are important considerations. A polymer with a relatively low viscosity (high melt index) and a fine grain size can be used to produce fine detailed rotational castings. A mesh sieve number 60 (250 μm) is a popular particle size for products less than 2 ft in diameter that have thin wall sections and undercuts. Larger particle sizes may be selected if simple designs or large parts are to be cast. (See Rotational and Centrifugal Casting.) Part wall thickness may be controlled by the speed and rotation of major and minor axes in rotational casting. Reversing the rotation during the heating cycle may help to fill voids in complicated molds (undercuts and sharp angles). This causes the powder and soft melt to change direction of flow. Rapid spinning causes reinforcements to compact and thus removes air from the mixture. Although there is a great deal of design freedom using rotational casting processing, tooling must be designed to compensate for material shrinkage. Wall thickness variations of less than 5% are difficult to control.

Decorative and furniture castings are commonly done in open-type tooling of metal, silicone elastomers, coated plasters, or polymers. (See Tooling.) Bathroom bowls, vanity and tabletops, ornamental plaques and filigree are familiar examples. Some are simple castings of resin and additives into molds, while others require gel coats followed by heavily filled and reinforced resins being poured into the mold. Very low pressures (5–10 psi) are used to fill some mold designs. With flexible molds, undercuts pose less of a problem in simple casting. Most designs should take into consideration the matrix shrinkage. Many elastomer molds must be supported or backed to assure quality parts. Liberal radii and draft angles are important tooling considerations. Like any process, the product finish is only as good as the mold. Therefore, releasing agents must be carefully selected to ensure proper part release and desirable surface finish.

Matched-die processing is a compression molding operation. The term **matched-die** has grown in use to indicate molding of large parts such as sanitary tubs, bathroom shower stalls, and numerous automobile panels from SMC. Premixes and preform compounds for traditional compression molding operations generally use short fibers or particulates. Matched-die molding of SMC allows molding of parts with longer fibers and a higher percentage of reinforcement. BMC, SMC, TMC, and preforms are matched-die (compression) molded. Heated metal molds are closed, forcing the material to flow throughout the cavity and forcing any entrapped air out of the reinforcing fiber or matrix. Fully positive molds are preferred with shear edges that seal in the resin and simultaneously shear off the reinforcing material. Many SMC operations are precut and require no mold pinch-off and produce no flash. Like traditional compression molding of granular materials or preforms, matched-die molding compounds should be preheated to produce high-quality parts and reduce the cycle time. (See Reinforcements and Molding Processes.) Parts with bosses, molded-in inserts, and ribs are possible using SMC in matched-die operations.

Radii, corners, and fillets should be liberal to reduce chipping, add strength, and simplify mold construction. As in most molding operations, it is best to avoid contour and stepped parting lines. Any variation of wall thickness should be a gradual transition. Any critical load areas or points of attachment where stress levels are expected to be high should be made thicker. Metal reinforcing or load spread plates may be molded-in or attached externally. It is best to produce parts molded from a single piece of reinforcement and matrix. Where layers or pieces of SMC must be used, make the overlapping bond area as large as possible because stress cracks generally develop across bond areas.

Preform matched-die molding operations generally require combining the reinforcing preform and the matrix ingredients at the press. Organic binders are commonly used to preshape reinforcements (preform) into a form similar to the shape to be molded. These preforms must be carefully designed to assure high-quality composite parts. (See Figure 6-17.) (See Reinforcements, Tooling.) In some methods, a gel coat is sprayed on the hot matched-die mold. After the gel has cured, the preform and matrix formulation are added and the molding cycle is completed. Many mold surfaces are chrome plated to provide die protection and improve the surface finish of the molded product. Mold release agents are commonly used. Design parameters are similar to other matched-die operations.

Reinforced RIM (RRIM) design parameters are relatively new but are similar to injection molding and structural foam designs.

The design parameters for **lay-up, spray-up, autoclave,** and **bag** techniques are similar. Except for reinforcement and matrix deposition, the processing characteristics are the same. If a high production rate is needed, it may require that more than one mold be produced since most open or contact molding operations have rel-

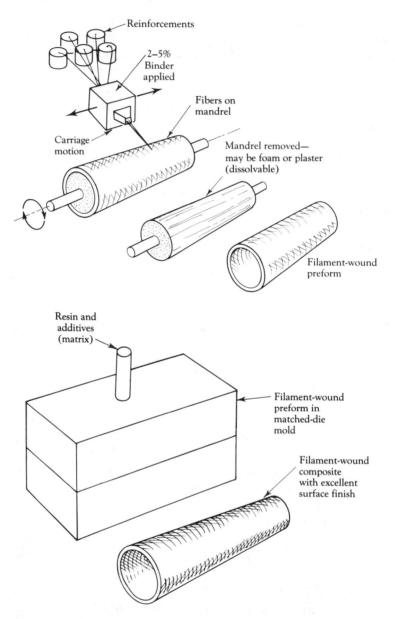

Figure 6-17. Polymer or plaster tooling may be used to make filament-wound preforms or finished parts. In this example the cellular (foam) mandrel is wound and solvents are used to remove the preform. The preform may then be placed in a matched-die mold for matrix injection and final cure.

atively long cycle times. Tooling is sometimes placed on wheeled carts to accommodate easy movement to different work stations such as production, finishing, and cleaning. Careful attention must be given to the matrix formulation and orientation of reinforcements. This may have more influence on the properties of the finished composite than the design. Composite properties are greatly influenced by the skill of the personnel performing the work. Preforms and SMC may be used in all operations. Tooling for lay-up and spray-up are essentially the same. Mold designs are either male or female, depending on which side must have a smooth surface. In all designs the reinforcing media should be lapped and not butted. Mat is commonly lapped a minimum of 1 in. Overlaps of 2 in. are preferred and they should be staggered. Designs should be provided with maximum radii and draft. Air bubbles may develop in areas with sharp angles. Zero-draft sections are possible in shallow designs but deeper designs may require split molds. Trim lines, location of holes, or attachment points may be inscribed in the mold surface to aid in fabrication. Simple, integrated part designs with gradual changes in thickness are desired. In highly stressed areas, additional thickness or metallic inserts may be required. Different glass-to-resin ratios or different formulations may be used in high-stress areas. Bosses and ribs are used for added strength but must be liberally tapered. Fins and hat sections are difficult and labor intensive to produce. Corrugating, dimpling, or crowning may provide adequate strength to large sheet areas. Molds must be designed with part removal in mind. Blow-out holes may be located in the bottom of the mold to assist in removal by pneumatic or mechanical means. Releasing agents are normally used but contamination of the composite surface for subsequent secondary bonding or coating operations must be considered.

Bag design parameters are similar to lay-up rules. Vacuum-bag, pressure-bag, and autoclave methods are used to densify, remove air bubbles, and remove excess resin. Prebleeding and compaction (debulking) is common in multilayered plies before curing. This requires additional cycle time to accommodate vacuum bagging and unbagging for the prebleeding and debulking. Most of these methods require considerable auxiliary equipment such as vacuum and curing chambers. In bag designs, a smooth finish is produced opposite the mold surface.

Reinforcement cutting (die, water jet, laser) and automated tape or fiber placement equipment may be required.

Pultrusion designs use long, continuous fibers or other reinforcing materials. Continuous strand mats, woven rovings, tapes, and cloth are used to provide some transverse strength. Although it is basically a closed-mold process, pultrusion imposes no limitations on draft angles. (See Pulforming.) Bosses, holes, raised numbers, or textured surfaces are not possible. Slots, holes, and grillwork can be cut, punched, or drilled after molding. As a rule, profiles should be kept simple with generous radii. The use of thermosetting polymers and heavy reinforcement results in little profile pull-back. Any fiber breakup reduces the reinforcing effects. Sharp corners or thickness transitions result in resin-rich zones with broken fibers. Ribs, undercuts, flanges, hollow sections, or corrugated designs require critical reinforcement placement and are possible only in the longitudinal direction.

Filament winding is another continuous reinforcement technique. The process, tooling, and equipment play a critical role in obtaining products with optimum properties. (See Molding Processes.) High structural efficiency may be achieved since the reinforcements can be oriented to match the direction and magnitude of stresses. In addition to reinforcement and matrix selection, the designer must consider the mandrel design and winding pattern that best meets all design criteria. Some mandrels are left in place to form a permanent liner or core for the filament-wound composite product. (See Tooling.) There are many options for removable mandrels. In either design (core or removable mandrels), considerable space may be required to store mandrels and the completed product.

Several excellent references are included in the appendix to provide detailed accounts of numerous winding patterns for optimal property designs. The winding process usually requires a computer to control the placement of filaments. These systems are designed to compensate for angle, contour reinforcing, bandwidth, equipment backlash, and other considerations in a precisely determined pattern. (See Figure 6-18.)

There are three basic winding patterns used to orient the filaments on the mandrel: (1) polar, (2) helical, and (3) hoop.

In polar winding (also called planar) the mandrel is wrapped (single-circuit pattern) with the filaments adjacent to each other. They are close to 0° to the longitudinal axis and there is no filament cross-

Figure 6-18. Computer-directed machine winds this composite tubular housing. Note the pattern of glass windings. (Engineering Technology Inc.)

over. The feed carriage may place several of these layers on the mandrel until the desired thickness and strength are achieved.

Helical winding is usually performed at angles of 15°–60° with the longitudinal axis. It is characterized by the filament crossover pattern on the mandrel. The optimum helix angle at which filaments are to be wound for a balanced design may be expressed as tan α = $\sqrt{2}$, or 54.75°. To determine the necessary number of repeating patterns, the total number of circuits per layer is expressed as

$$C_L = \frac{D}{S_c}$$

where

C_L = circuits per layer
D = mean composite layer
S_c = circumferential component of bandwidth

A multiple (ten) circuit helical pattern is shown in Figure 6-19. The contours of the head region are most difficult and critical. Filaments must be carefully controlled and oriented. Wet filaments tend to slip or change position as they are wound under tension. Winding tension varies but is generally less than 1 lb per filament end.

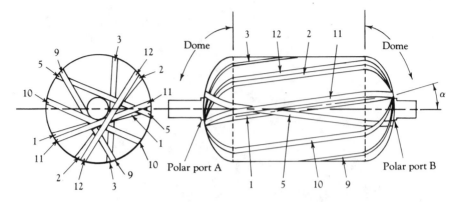

Figure 6-19. Multiple (ten) circuit helical pattern. (E.C. Young, 18th Petroleum Mechanical Engineering Conference)

Hoop or circumferential winding places the filaments close to 90° to the longitudinal axis. Several layers may be applied to increase strength. Sometimes hoop layers are placed on other design patterns as localized stiffeners.

Bosses, fins, and undercuts are not possible with conventional techniques. Because of winding and fiber defects, only about 70% of the ultimate strength may be realized in most filament-wound designs.

Translucent panels are made in a variety of configurations and colors. They are commonly made of fiber lengths of less than 2 in. Some use continuous filaments and woven fabric. As a rule, there are irregularities in the panel appearance caused by uneven or random placement of reinforcements. The design parameters are generally simple, depending on the processing technique. Hand lay-up is generally the least expensive. Sheet metal plates are used as the molds. The process, called the batch or stacked-panel process, allows several layers of molds to be stacked.

Heated platens are sometimes used to press composite formulations into textured molds. Press processing provides improved thickness control over lay-up methods.

Continuous processing is the most efficient, high-production method for producing composite flat, corrugated, or other panel configurations. Nip-roll and tractor-type tooling are costly.

Laminar Composite Design Guidelines

Laminar composites consist of layers (lamina) of at least two different materials bonded together. (See Molding Processes, Laminating/Laminated.) These layers may differ in material, form, and/or orientation. Some layers may themselves be composites.

Laminated packaging materials have become a major factor in the distribution of products throughout the world. These composite laminates vary greatly. Paper, metal foils, and numerous polymers are bonded together to form unique and innovative products. The selection of each ply layer is of primary importance if the product must have a high-barrier to odors, fragrances, oxygen, and water vapor. Packaging foodstuffs requires FDA approved polymers. Sometimes the outer layer must be sealable by impulse bonding and easily labeled. (See Film Laminates, Extrusion, and Coextrusion.)

Because of the great variety of possible combinations of layers, materials, forms, and orientation, it is difficult to generalize about laminar designs. In theory, there are as many composite laminates as there are possible combinations of two or more materials. The unique properties obtained from layered construction allow the designer to develop composite materials to meet new and innovative design parameters. Products may be composed of metallic flakes in matrix bonding layers of glass and organic cloth. This composite laminate may be bonded to one side of a foam-filled honeycomb core, and a different skin material or composite may be bonded to the other side. The resulting product may be used as a light, strong, composite panel for protection against lightning and electromagnetic interference.

It should be apparent that this definition allows for some confusion and crossover of processing. Some products produced from spray-up, lay-up, and vacuum-bag molding techniques may qualify as laminar composites. For example, the spray-up of fibrous glass and polymer matrix into a thermoformed thermoplastic shell (a process called rigidizing) results in a laminar composite. This technique is normally classed as a reinforcing process. Hand lay-up of different plies, orientation, and materials also qualifies as a laminate.

Combinations of matrix, fibrous mat, cloth, or other materials may be "laminated" together in a continuous processing technique. The definition of laminates becomes more complex with the inclusion of hybrid laminates. Hybrid forms are combinations of different reinforcing and/or matrix materials. The form may consist of short, mixed-glass fibers in a matrix something like SMC and layers of alternating continuous graphite fiber and glass layers. Hybrid forms may be processed by compression and some bag molding techniques. Some pultrusion and filament-winding systems with alternating orientations of two different reinforcements may be considered laminates. Copper-clad laminates with a phenolic matrix and layers of cloth, paper, or fibrous glass are familiar hybrid laminates used for electrical circuit boards. Decorative laminates used on kitchen countertops and furniture surfaces are a direct result of developments in laminated circuit boards.

When two or more plies of unidirectional composites are stacked in a particular order, the composite hybrid is called an **interply hybrid.** When two or more different fibers are in the same ply, it is known as an **intraply hybrid.** A composite consisting of matrix, fibers, and metallic plies is called a **superhybrid.** Aircraft turbine engine blades made from plies of graphite, glass, Kevlar, boron, aluminum, and an epoxy matrix are superhybrids.

Sandwich composites are laminates with a thick low-density core and thin high-density faces (skins). In all laminar designs, the role of the interface (bonding energies) is of primary importance. There must be a strong bond between the matrix and reinforcement plies. Surface preparation is very important. Rough scratches may affect bond strength if the peaks are sharp-edged. They may act as stress-concentration points. The properties of different constituents, such as thermal expansion and elasticity, may result in locked-in stresses or result in shear stresses. Designers may use the directional shear properties of hexagonal honeycomb cores so that the major shear stresses are carried in the longitudinal ribbon direction.

To minimize weight, adhesive or matrix compounds are not simply spread over the skin sheets. Rollers are commonly used to apply the bonding agent only on the edges of the honeycomb structure. There is little bridging action between cells. (See Figure 6-20.) It does little good to apply adhesive where no bond is intended. A transfer film (Tedlar) is sometimes used to apply an even coat to the core. A film is coated with adhesive. After the coated film is rubbed against the core, the film is removed, leaving an adhesive coat on the core.

Stress in each layer is proportional to the distance from the neu-

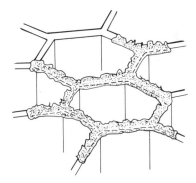

Figure 6-20. Bonding agents coat edges of honeycomb structure prior to bonding.

tral axis and the moduli of elasticity of the layers. This concept is shown in Figure 6-21.

The primary purpose of sandwich composites is to achieve high strength with less weight (high strength-to-weight ratio). High-density facings correspond to the basic principle of an I-beam. These flange facings act together to resist most of the applied and bending loads. The lightweight core must be stiff enough to resist transverse ten-

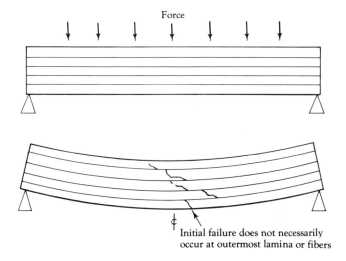

Force

Initial failure does not necessarily occur at outermost lamina or fibers

Figure 6-21. Bending of laminate induces internal stresses proportional to moduli of elasticity and distance from the neutral axis.

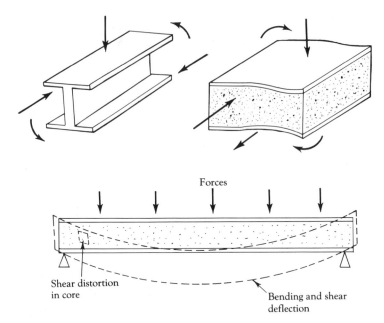

Forces

Shear distortion
in core

Bending and shear
deflection

Figure 6-22. Facings on sandwich composites are similar to flanges of I-beam section. They must resist transverse tension, compression, shear, and buckling.

sion, compression, shear, and buckling. This concept is shown in Figure 6-22.

The longitudinal or 0° values cited in the literature are not very useful in understanding the multidirectional properties of composite laminates. Laminating processes may also influence and greatly reduce these values. Many of the property-reducing factors, such as stress concentrations, fiber wash, matrix-reinforcement bond, wetting, resin-rich areas, and voids, are determined as much by fabrication as by properties of the component materials.

The tensile moduli of composites are greatly reduced as the angle between the fibers and the direction of tensile load is increased. The failure modes of single and multidirectional fiber orientation are shown in Figures 6-23 and 6-24.

Balance and stack symmetry must be maintained to avoid distortion. Balance implies that there must be a −45° ply for every +45° ply in the laminate. Unusual deformations as a result of simple loads are shown in Figure 6-25. A design close to optimum and that

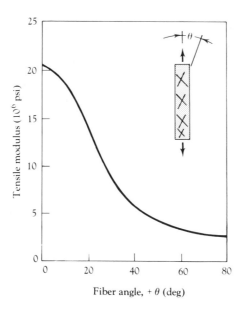

Figure 6-23. Tensile modulus of carbon/epoxy composites drops steeply as the angle between the fibers and the direction of tensile load is increased. (With permission, *Machine Design*)

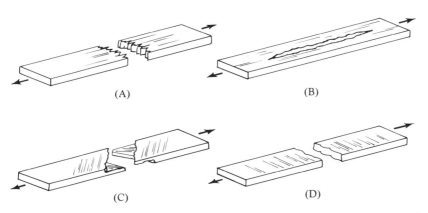

Figure 6-24. Failure modes of composites in tension. Tensional strength of a carbon/epoxy structural composite is always related to fiber direction. A simple tensile test shows strikingly different failure behavior in composites having different fiber orientation. In a multidirectional composite, single plies can fail without overall structural failure. Recognition of the various failure modes and knowing how composites fail are prerequisites to determining a fix. (With permission, *Machine Design*)

resists all loads may be a laminate consisting of plies at 0°, ±45°, and 90°.

Typical charts from the *Advanced Composites Design Guide* show the modulus and strength levels for the 0°/±45°/90° family of high-strength composites containing 65% (by volume) carbon fibers in an epoxy matrix. The principal design criterion is concerned with fiber orientation in each layer. Fibers should be oriented in the direction of the primary stresses. Fibers arranged in a random manner (isotropic) have equal strength in all directions.

The objective of the designer is to orient the lamina in the principal direction of anticipated stresses. Since each layer of the laminate will attempt to deform independently, a mathematical means may be used to determine the composite laminate properties. There

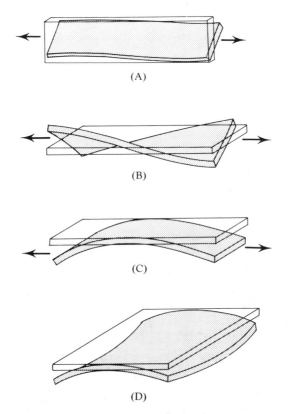

Figure 6-25. Symmetry effects on deflection of composites. (With permission, *Machine Design*)

are several models and computer programs cited in the literature to help determine the combined properties of all layers using different materials and fiber orientation. Unidirectional and multidirectional properties of selected materials are shown in Table 6-4.

Automated tape and fiber placement allow for a direct lay-up on a flat or contoured tooling. The result is a more uniform, repetitive placement at higher processing speeds. The multiaxis motion of the delivery head can be programmed to place reinforcement forms in many configurations.

Composite joint designs must consider the advantages and disadvantages of mechanical fasteners and adhesive bonding. These are summarized in Table 6-5. The actual joining design must take into consideration the anisotropic characteristic of the composite laminate. In good joint designs, the surface fiber plies are bonded in the direction of or parallel to the load. Scarf and landed-scarf joint configurations are acceptable designs. These concepts are shown in Figure 6-26.

Mechanical fasteners may be needed for added strength or disassembly. Drilled holes are preferred to molded-in holes. Several methods may be used to prevent shear and cleavage cracks. Do not reinforce a hole or cutout with carbon fibers aligned in the direction of the load. It is best to cross-ply fibers at 45° or 90° to the load. Metal shims and epoxy adhesives are sometimes used to fill any clearance between the fastener and the hole. Both treatments help to distribute load stresses. Bonded-mechanical fastener joints are superior to either adhesive or mechanical fastened joints.

Design criteria for hole diameter and composite thickness are expressed as D/t (hole diameter divided by composite thickness). For boron and graphite composites the D/t ratio should be 3. Conservative designs may use a D/t ratio of 1.

Joints for sandwich construction pose a special problem. Scarf joints are low cost and attractive but have little strength. Densifying the edge or crushing the cellular core are sometimes used for edge treatments and joint preparation. The strongest joints are produced by using external or internal inserts and doublers where needed. Many facings must be made strong enough to withstand the compressive and shear stresses induced on mechanical fasteners by the loads. Specially designed extruded or fabricated doublers are commonly used.

Table 6-4. Unidirectional and Multidirectional Properties of Selected Materials[a]

Material[b]	Density (lb/in.³)	Unidirectional Properties						±45° Properties			
		Modulus (10⁶ lb/in.²)			Ultimate Strength (kips/in.²)			Modulus (10⁶ lb/in.²)		Ultimate Strength (kips/in.²)	
		Axial E_{11}	Trans E_{22}	Shear G_{12}	Axial σ_{11}	Trans σ_{22}	Shear τ_{12}	$E_x = E_y$	G_{xy}	τ_{xy}	$\sigma_x = \sigma_y$
C-Ep											
High strength	0.057	20	1.0	0.65	220	6	14	2.5	4.5	20	50
High modulus	0.058	29	1.0	0.70	175	5	10	2.5	6.5	18	42
Ultrahigh modulus	0.061	44	1.0	0.95	110	4	7	3.0	11.5	14	30

| Material[b] | Density (lb/in.³) | Unidirectional Properties | | | | | | ±45° Properties | | | |
| | | Modulus (10⁶ lb/in.²) | | | Ultimate Strength (kips/in.²) | | | Modulus (10⁶ lb/in.²) | | Ultimate Strength (kips/in.²) | |
		Axial E_{11}	Trans E_{22}	Shear G_{12}	Axial σ_{11}	Trans σ_{22}	Shear τ_{12}	$E_x = E_y$	G_{xy}	τ_{xy}	$\sigma_x = \sigma_y$
Kv 49-Ep	0.050	12.5	0.8	0.3	220	4	6	1.1	3.0	30	32
EGI-Ep	0.072	6	1.5	0.3	180	6	10	1.5	3.0	25	40
Chopped-glass SMC											
30% by volume	0.068	2.5	2.5	1.0	30	30	20	2.5	1.0	20	30
60% by volume	0.072	3.5	3.5	1.5	50	50	40	3.5	1.5	40	50
Steel	0.294	29.5	29.5	11.5	60	60	35	29.5	11.5	35	60
Aluminum	0.098	10.5	10.5	3.8	42	42	28	10.5	3.8	28	42

[a]Reproduced with permission from *Machine Design*, Dec. 6, 1979, pp. 150 ff.
[b]Fiber content is 65 vol. % except for steel and aluminum, where it does not apply, and for chopped-glass SMC, where fiber content is 30 vol. % and 60 vol. %.

Table 6-5. Advantages and Disadvantages
of Adhesive Bonds and Mechanical Fasteners

Advantages	Disadvantages
Adhesive Bonds	
Light weight	Thermal cycling and degradation
Distribution of load over joint	Humidity and chemicals weaken
No holes needed	bonds
Many joint designs require no	Requires surface preparation
machining	Joint disassembly and
	inspection not possible
	Low shear strength
	Adhesive thermal service limit
Mechanical Fasteners	
Disassembly permitted	Added weight and bulk
Joint inspection possible	Holes must be cut
Little surface preparation	Plies are weakened
needed	Stress concentrations at holes
Thermal service limited to	Corrosion with graphite
matrix	Low bearing strength
Low facility and tooling	Generally higher cost
requirements	

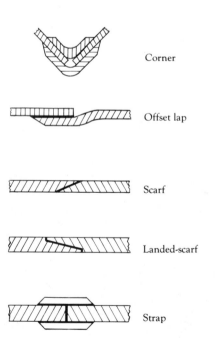

Corner

Offset lap

Scarf

Landed-scarf

Strap

Figure 6-26. Multilayered composite joint designs.

Particulate Design Guidelines

Remember that the term **particulate** has been limited to those particles with high aspect ratios. Most fillers do not qualify because they are not platelike in nature. Particulates, however, may serve as fillers or extenders. (See Fillers, Reinforcements, and Particulates.)

One of the problems with glass- or fiber-reinforced polymers is that shrinkage in the flow direction is less than in the transverse direction. Fibers tend to orient in the flow direction. Because of the platelike nature of particulates, shrinkage is uniform in both flow and transverse directions. Remember, particles have more uniform mechanical properties (isotropic) than fibers. (See Reinforcements, Particulates.) Because of the flake or platelike shape of the particulates, they are able to overlap and pack tightly together with few voids in the matrix. Maximum strength is perpendicular to the flake plane. Particulate composites have shear strength equal to the matrix. Flake composites can have a higher theoretical modulus than fiber composites.

The design parameters for particulate composites are similar to fibers and fillers. The effects of particle size, size distribution, shape, surface treatment, and type of particle will change flow behavior and product properties. Designs and processing techniques should be provided to allow parallel orientation of flakes in the part. Potentially, the most important variable is maintaining the aspect ratio and parallel packing of the particulate. Processing may break or distribute the particles in random fashion. Metallic flakes are used in EMI shielding applications or where electrical conductivity is needed. (See Figure 6-27.)

Machining Techniques

Machining techniques for composite materials attempt to prevent fraying and delamination of the composite. There are a few broad criteria that must be met. The abrasive nature of most composites requires that cutting tools be made from tungsten carbide or be coated (e.g., with titanium diboride) when possible. High-speed steel (M2) or diamond-tipped cutters are also used. Diamond tools are preferred in cutting boron/epoxy. Water-jet and laser cutting systems are clean processes producing finished cuts on most uncured materials. (See Figure 6-28.) Shearing and punching operations are similar to sheet metal die designs. Abrasive jet and ultrasonically

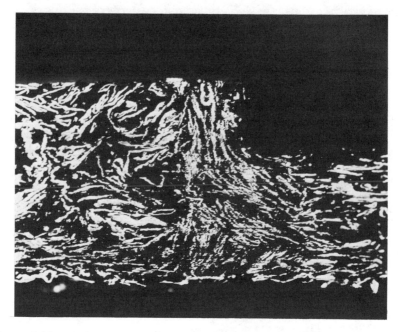

Figure 6-27. Cross section of an injection molded part shows typical distribution and orientation of aluminum flakes. Conductivity of particulates provides EMI shielding. (Transmet Corporation)

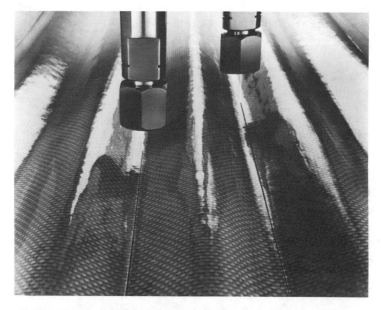

Figure 6-28. Space-age composites like Kevlar are quickly and cleanly cut by water jet. (Flow Systems, Inc.)

vibrated cutting tools may be used to cut round or odd-shaped holes in most composites. In Figure 6-29 a water-jet robotic system is used to degate and deflash automotive front end panels.

As a general rule, cutters must be sharp and smooth and require little clearance. Clearance angles of about 5°–10° and point angles of 60°–120° are recommended for most drill bits. Zero rake is recommended on most circular saw blades. Bandsaw blades may be reversed and accomplish a zero or negative rake. Abrasive carbide or diamond grit blades may be used to cut graphite and boron/epoxy composites. It is best to back up any cutting or drilling operation with a solid material to reduce chipping, fraying, and delamination.

The feeds and speeds for cutting vary greatly with thickness and material but are similar to those for nonferrous materials such as aluminum. A vacuum system should be used to remove hazardous materials from the environment. Liquid lubricants and coolants are used to prevent matrix thermal degradation and protect personnel. (See Table 6-6.)

Standards, Specifications, and Sources of Help

The term **standard** can have different meanings. It can be a written set of technical, dimensional, or performance requirements; or

Figure 6-29. Robotic-directed water jet used in degating and deflashing automotive parts. (Sterling–Detroit Company)

Table 6-6. Machining Composites

Operation	Material	Cutting Tool	Speeds	Feeds (<0.250 in. thick)
Drilling	G1-Pe	0.250-Diamond	20,000 RPM	
	B-Ep	0.250-Diamond core (60–120 grit)	100 SFPM	0.002/rev
				0.002/rev
	B-Ep	0.250 2-4 Flute (HSS)	25 SFPM	0.002/rev
	Kv-Ep	Spade–carbide	>25,000 RPM	
	Kv-Ep	Brad point–carbide	>6,000 RPM	
	G1-Ep	Tungsten carbide	<2,000 RPM	<0.5 ipm
	Gr-Ep	Tungsten carbide	>5,000 RPM	<0.5 ipm
Bandsaw	Kv-Ep	14 Teeth, honed saber	3,000–6,000 SFPM	<30 ipm
	B-Ep	Carbide or 60 gut diamond	2,000–5,000 SFPM	<30 ipm
			3,000–6,000 SFPM	<30 ipm
	Hybrids G1-Pe	14 Teeth, honed saber	3,000–6,000 SFPM	<30 ipm
Milling	Most	Four-flute carbide	300– 800 SFPM	<10 ipm
Circular saw	Gr-Ep, B-Ep	60 Grit diamond		
	G1-Pe	60 Tooth carbide or 60 grit diamond	6,000 SFPM	<30 ipm

Process	Material	Tooling	Speed	Feed
Lathe	Kv-Ep Gl-Pe	Carbide Carbide	250–300 SFPM 300–600 SFPM	0.002/rev 0.002/rev
Shears	Kv-Ep	—	—	<30 ipm
Countersink or counterbore	Most	Diamond grit or carbide	20,000 RPM 6,000 RPM	<0.5 ipm
Laser (10 kW)	Most <0.250 in. thickness	CO_2 cooling	—	<30 ipm, depending on material
Water-jet	Most <0.250 in. thickness	60,000 psi, 0.10 in. orifice	—	<30 ipm, depending on material
Abrasive (sanding–grinding)	Most	Silicone carbide or alumina grit (wet)	4,000 SFPM	—
Router	Most	Carbide or diamond grit	20,000 RPM	—

it can be an accepted process or procedure. Standards are generally used not to describe products and services but to establish requirements of meeting safety, environmental protection, welfare, and other national objectives.

There are a number of different types of standards including **physical** standards like those kept by the NBS, **regulatory** standards such as those from the EPA, **voluntary** standards recommended by technical societies, producers, trade associations, or other groups such as UL, and **public** standards promoted by government bodies and professional organizations such as ASTM. Sometimes **private** standards are developed by companies when other standards are lacking or inadequate. The majority of all standards originate from industry. Many of these are freely adopted (voluntary) by other industries or standard making organizations.

Metrication and internationalization of standards have direct and indirect impacts on economic considerations. The metric system is an important part of the international measuring system. The U.S. customary measurement system is the most complicated one invented. It involves time-consuming calculations and different conversions. We continue to place on the world market the only manufactured products using an outmoded measurement system. It seems strange that a nation with outstanding technological achievement, trying to sell its products throughout the world, would cling to a measuring system based on ancient, nostalgic standards.

Engineering standards are not the same as measuring standards. These standards are used to specify how big, heavy, hard, and so on something is, not how it is measured. Engineering standards specify how wide and tall an exterior door is, not the measurement standards. We could measure the width and height of the door in inches or millimeters and it would not change the physical standard. We must change our attitude about international standards if we are to increase our international trade significantly.

Savings from adopting world standards may include reduced costs associated with materials, production, inventory, design, testing, engineering, documentation, and quality control. There is substantial evidence that increased standardization lowers costs, but it also restricts choice. It is doubtful that standardization will inhibit innovative or technological ideas. Standards change and there are allowable variations in product quality and performance.

The *Department of Defense Index of Specifications and Stand-*

ards (DODISS) is a reference publication made available to private industry in microfiche or printed book format. The DODISS listings include the following unclassified documents:

Military Specifications and Military Standards
Federal Specifications and Federal Standards
Military Handbooks
Qualified Products Lists and Industry Documents
Air Force–Navy Aeronautical Standards and Aeronautical Design Standards
U.S. Air Force Specifications and Specification Bulletins
Air Force–Navy Aeronautical Bulletins

Additional information and help may be obtained from the following sources (also see Sources of Safety Information):

American National Standards Institute (ANSI)
1430 Broadway
New York, NY 10018

American Petroleum Institute
1801 K Street, NW
Washington, DC 20006

(The) American Society for Testing and Materials (ASTM)
1916 Race Street
Philadelphia, PA 19103

(The) American Society of Mechanical Engineers (ASME)
United Engineering Center
345 East 47th Street
New York, NY 10017

Defense Standardization Program Office (DSPO)
5203 Leesburg Pike, Suite 1403
Falls Church, VA 22041-3466

Department of Defense (DOD)
Office for Research and Engineering
Washington, DC 20301

General Services Administration (GSA)
Federal Supply Service
18th and F Streets
Washington, DC 20406

Global Engineering Documentation Services, Inc.
3301 W. MacArthur Boulevard
Santa Ana, CA 92704

Instrument Society of America
400 Stanwix Street
Pittsburgh, PA 15222

International Organization for Standardization (ISO)
1 rue de Varembe,
CH 1211
Geneve 20 Switzerland/Suisse

National Bureau of Standards (NBS)
Standards Information & Analysis Section
Standards Information Service (SIS)
Building 225, Room B 162
Washington, DC 20234

National Conference on Weights and Measures
c/o National Bureau of Standards
Washington, DC 20234

Navy Publications and Printing Service Office
700 Robbins Avenue
Philadelphia, PA 19111

Scientific Apparatus Makers Association
1140 Connecticut Avenue, NW
Washington, DC 20036

Society of Automotive Engineers (SAE)
400 Commonwealth Drive
Warrendale, PA 15096

Society of Plastics Engineers, Inc. (SPE)
14 Fairfield Drive
Brookfield Center, CT 06805

Society of Plastics Industry, Inc. (SPI)
1025 Connecticut Avenue, NW, Suite 409
Washington, DC 20036

Underwriters Laboratories (UL)
207 East Ohio Street
Chicago, IL 60611

U.S. Metric Association
Sugarloaf Star Route
Boulder, CO 80302

Performance Testing

Testing techniques and data obtained for composite materials with discrete particles, laminar layers, or continuous fibers differ greatly in many respects. Some are quasi-isotropic, while others are essentially anisotropic. A specimen cut from metal is generally considered to represent the properties of that metal closely, with a general relationship between these properties and those of the final product. Unlike metals, composites may have numerous types of defects such as voids, delaminations, debonds, wrinkles, moisture, inclusions, broken or damaged reinforcements, reinforcement misalignment, and matrix cracking. (See Table 6-7.) Because many of these defects may be present, designers must use caution in drawing meaningful conclusions about a particular composite test sample. Even the test method or procedure may provide erroneous assessment data. The test method may be dependent on materials used, design, and processing. No single method is generally applicable to all composite parts.

There has been some confusion about the terms **inspection, quality control**, and **testing.** The term **inspection** ensures that during manufacture of a composite part personnel conduct visual examinations of materials, placement of plies, orientation of reinforcements, gauge readings, and so on. Any flaws in processing or materials must be detected. The term **quality control** is a procedure to determine if a product is being manufactured to specifications. Although inspection and quality control are often used interchangeably, quality control is a technique of management for achieving quality. Inspection is part of that technique. The term **testing** implies that methods or procedures are used to determine physical, mechanical, chemical, optical, electrical, or other properties of a composite part.

These methods are sometimes divided into destructive or nondestructive testing. No single method is able to detect all defects. In all destructive testing procedures, the sample is damaged or destroyed. Many of these tests are published by various agencies including ASTM. Nondestructive tests are designed to obtain data without damaging the composite part or sample. Nondestructive

Table 6-7. Selected Composite Defects and Possible Causes

Defect	Possible Causes
Voids	Moisture; air; poor resin flow; improper winding tension, pressure, or reinforcement placement; need to degas, improper processing
Delaminations	Thermal or residual stress; improper winding tension, pressure, or reinforcement placement; mismatch in thermal coefficient of expansion; moisture, improper processing, subquality materials, improper design
Debonds	Excessive resin; improper surface treatment and cure cycle; moisture contaminants; improper processing; fabrication environment; subquality materials; improper design
Wrinkles	Improper tension, alignment, placement, and pressure; trapped air, need to outgas, improper design
Moisture	Improper drying and preparation accelerate degradation and reduce adhesive bonding
Inclusions	Any unwanted material, such as dirt, loose fibers, and lubricants, resulting in general reduction in selected properties; fabrication environment
Damaged reinforcements	Cut or broken fibers; improper tension, processing, or handling; subquality materials
Reinforcement misalignment	Improper processing; poor inspection; improper design parameters, winding tension, pressure, or handling; improper design
Matrix cracking	Molded-in stresses; resin-rich areas; improper cure and processing; subquality materials

testing methods may include radiography (x-ray), ultrasonics, acoustic emission, thermography, holography, computer tomography, penetration, microwave, temperature differential, and infrared scanning. (See Table 6-8.)

Most of these tests are performed on specified samples under controlled laboratory conditions. In **tag end** testing, specimens are cut from the actual part. The specimen must be representative of the whole part to provide some degree of reliability. **Verification** test panels are also used but they may not be representative of the actual part. **Proof** testing may be required in some critical design areas. A composite tank, rocket case, or pole-vaulter's pole may be subjected to a pass/fail test (proof). Although the part may pass the test, proof testing does not provide information about any defects present in the structure which may lead to further failures.

Tests can only be used as indicators for product services and design. Testing limitations should be kept in mind when making judgments about composites. The criticality of a defect may depend on design requirements. Some defects may have only a limited effect on the performance of the part. To ensure product success, the service requirements, design, and properties of the composite product must be carefully considered. The true test for any product is the performance and use under actual service conditions.

Product Applications

The potential and future of polymer composites looks bright. It has been predicted by a Stamford-based study (1986) that composite parts will grow 15% per year over the next 10 years.

We have begun to design composite products for "manufacturability" and "automation." Most composites have a good balance of physical properties, excellent strength-to-weight ratios, corrosion resistance, good electrical properties, and low tooling cost. It is lower material costs, ease of fabrication, integrated part designs, faster molding cycles, improved finishing, product reliability, environmental impact, and energy savings that will accelerate the growing consumer acceptance of composites.

Aerospace and military research has changed our concept of design and traditional construction techniques. It took nearly 20 years after the all-aluminum Junkers aircraft flew (1922) before all-metal aircraft became common. Wood and fabric were major materials in

Table 6-8. Selected Techniques to Detect Defects

Defects Detected	Radiography	Computer Tomography	Ultrasonics	Acoustic Emission	Acoustic Ultrasonics	Thermography	Optical Holography	Eddy Current
Voids	Yes	Yes	Yes	No	Yes	Some	Yes	Yes
Debonds	Some	Yes	Yes	Yes	Yes	Some	Yes	Yes
Delaminations	Some	Yes	Yes	Yes	Yes	Some	Yes	Yes
Impact damage	Some	Some	Yes	Yes	Yes	Some	Yes	Yes
Density variations	Yes	Yes	Yes	No	Yes	No	No	Yes
Resin variations	Yes	Yes	Yes	No	Yes	No	No	Yes
Broken fibers	Some	Some	Yes	Yes	Yes	No	Yes	Yes
Fiber misalignment	Yes	Yes	Yes	Yes	Yes	No	No	Yes
Wrinkles	Yes	Yes	Yes	Yes	Yes	Yes	Yes	Yes
Resin cracks	Some	Yes	Yes	No	Yes	Yes	Yes	Yes
Porosity	Yes	Yes	Yes	No	Yes	Some	Yes	Yes
Cure variations	No	Yes	Yes	No	No	No	No	Yes
Inclusions	Yes	Yes	Yes	Yes	Yes	No	No	Yes
Moisture	No	Yes	Yes	No	Yes	Yes	Yes	Yes

aircraft and other transportation systems at the time. The largest growth potential for composite materials is in the areas of transportation and construction.

Automobiles, trucks, buses, vans, rapid transit vehicles, aircraft, boats, ships, aerospace vehicles, and trains will benefit from composite design. (See Figure 6-30.) The spin-off from space and military program research has allowed the civilian markets to make impressive gains. If the federal regulations of the National Highway and Transportation Safety Administration's (NHTSA) proposed Corporate Average Fuel Economy (CAFE) of 40 mpg is realized by the year 2000, automobiles will continue to be downsized. (American-built and -engineered automobiles will cost less than foreign models.) Ford, General Motors, and Chrysler have developed all-composite concept automobiles, but according to one manufacturer, the all-composite automobile will not appear before the year 2000. By the year 2000, one-half of all cars could have composite bodies. Specifically GM's 1988 A-van, the 1990 Firebird and Camaro, and the 1992 J-car will have composite bodies. Chrysler plans an all-composite, low-cost, high-volume car called Genesis. In 1986, a typical automobile has some 15,000 parts. Composites could reduce this number to hundreds. In the meantime, the list of automotive components

Figure 6-30.   Trains and mass transit vehicles will use composite materials and designs.

will grow. Present production applications include driveshafts, side rails, doors, cross members, oil pans, suspension arms, leaf springs, wheels, quarter panels, trunk decks, hoods, hinges, transmission support, front and rear bumper components, seat frame, and wheels. Many engine components are being tested in present racing engines. (See Figure 6-31.)

Composites are being developed for a variety of highly mobile, easily transported vehicles for tactical, ballistic (armor) combat and logistical support roles.

Composite materials, in addition to providing weight savings over the life of a vehicle, are also energy efficient. It has been estimated that for every 100 lb reduced from a 2500 lb automobile, there will be a fuel savings of 0.3 mpg. The energy resource requirements per pound of most composite materials are only half that of aluminum

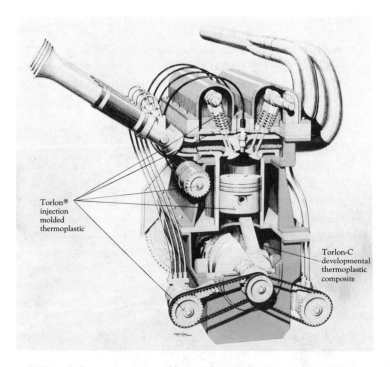

Torlon®
injection
molded
thermoplastic

Torlon-C
developmental
thermoplastic
composite

Figure 6-31. Schematic view of an engine includes piston skirts, piston pins, piston rings, connection rods, valve stems, engine block, and cylinder heads made of carbon fiber–polyamide–imide composite. (Amoco Chemicals Company)

or steel. Competition for raw material supplies between energy and petrochemical uses will continue to be a concern.

New truck designs will incorporate light, strong, aerodynamic bodies and cargo containers. Reduction in wind resistance and vehicle weight will allow room for more cargo with less fuel consumption. Many of these concepts will be used for buses, trains, and mass transportation systems.

The search for higher strength-to-weight ratios and design flexibility has widened in aircraft and aerospace designs. Glass fiber composites have a specific strength five times that of aluminum, while graphite and boron in epoxy matrix have about a five times higher specific stiffness.

The current list of composite aircraft and aerospace applications is impressive. The all-composite horizontal stabilizer for the B-1 bomber, vertical and horizontal stabilizers on the F-16 fighter, wing skins on the F-18, rudder and flaps for the A-4, helicopter rotor blades, rudders, elevators, ailerons, and spoilers on the B767 and B757, and payload doors of the Space Shuttle Orbiter are the most familiar applications. It has been estimated that 20% of the total weight of the AV-88 (Harrier) will be composed of polymer composites. Even greater projected use of composites are in the future for AFT, V/STOL, HOTOL, Stealth bombers, and Mach 20 transports. It has been suggested that up to 65% of the structure weight may be composed of polymer composites. (See Figure 6-32.)

Glass fiber composite pleasure boats have been used for years. The 153 ft H.M.S. Wilton in the British Navy is the world's largest composite ship. It is likely that continued production of small ships in the minesweeper, hydrofoil, and surface effect ship class will continue to grow. Corrosion resistance, light weight, and low maintenance make marine use in hull structures, fairwaters, sonar domes, antennas, floats, buoys, masts, spars, deckhouses, fairings, and all kinds of tanks an attractive application. Carbon composites are very radar absorptive and are likely to find additional structural applications in both aircraft and ships.

The market penetration in sporting goods is expected to grow from 40 to 50% in the next 5 years, replacing wood and metal. Graphite, boron, and Kevlar in an epoxy matrix have found acceptance in golf carts, surf boards, hang-glider frames, javelins, hockey sticks, sailplanes, sailboats, ski poles, safe playground equipment, golf shafts, fishing rods, snow and water skis, bows, arrows, tennis

Figure 6-32.   These F-16 fighters are composed of many composite parts. (General Dynamics)

rackets, pole-vaulting poles, skateboards, bats, helmets, bicycle frames, canoes, catamarans, oars, paddles, and other items.

With changes in building codes, craft unions, consumer acceptance and acceptable designs, construction has potential for considerable growth in composite usage. Not everyone can afford a custom-built home made by traditional methods. Mass-produced, modular manufactured composite homes may be in our future. The growing mobile-home, apartment, hotel, and motel market is a natural for modular composite construction.

There are a number of miscellaneous applications, including soundboards for guitars and violins, shells for musical drums, and lightweight armored products for personnel, vehicle, and equipment protection. The appliance and furniture industries will continue to initiate new designs using the merits of composites. Housings, frames, bases, tanks, and fans will be used in the manufacture of computers, vacuum cleaners, air conditioners, refrigerators, and other appliances. Chairs, lamps, tables, and other modern furniture designs are destined to be made of composite materials. Biocompatible implants, prostheses, electrical circuit boards, hammer handles, ladder rails, highway signs, wheel chairs, and numerous pipes and ducting for the food and chemical processing industries are present applications.

The proliferation of new products are announced daily. With additional research and innovative designs, there will be an unprecedented demand for composite products.

Chapter 7. What Are the Safety and Hygiene Hazards?

Introduction

There are a number of safety and health problems and concerns unique to the plastics and petrochemical industries. Critics attack these industries for their economic and environmental actions as well as their safety and occupational health policies and programs.

Charles Kettering used to say, "Chance favors the prepared mind." Similarly, success favors those who understand and take the time to educate the workforce and public. Responsible designers and managers must consider the safety of the public and personnel operating processing equipment. All hazards should be considered, and an environmental impact statement should be made into a document. This document could then be used to consider acceptable design approaches for safety.

A safety education program may consider access to equipment, fire protection, toxic chemical hazards, sewer problems, or other potentially hazardous scenarios. Although such a document is neither required nor justified in all industries, it is a valuable instrument in helping to avoid future safety problems and providing factual information for all those concerned. (See Safety Factors.)

Chemical manufacturers are required to develop a Material Safety Data Sheet (MSDS) for all hazardous chemicals, clearly place product labels on all containers, and provide this information to customers. Since 1986, employers (Standard Industrial Classifications Code 20-39) must train employees, label all in-plant containers, and have a chemical list, MSDS file, and a complete written program.

OSHA does require a log of occupational injuries and illnesses. Employers have the responsibility (under OSHA's Hazard Communication Standard) to provide employees with information about chemicals they handle and use on the job. It is advisable to document the names of workers who have received training and which chemicals are covered. Record all requests for information.

Personnel Safety

Hazards associated with the production of composite materials may be divided into three major areas: (1) chemical, (2) physical, and (3) biomechanical.

Chemical hazards are found in the form of liquids, mists, vapors, gases, and dusts. They may be ingested, inhaled, or absorbed. Some may be irritants, narcotics, or carcinogens. Organic solvents, fumes, and unreacted polymers pose most of the danger. Most fully reacted (cured) polymers are inert and nontoxic. Many are biocompatible. Plasticizers, solvents, lubricants, stabilizers, colors, and other additives may leach out of the polymer matrix and come in contact with the human body or be absorbed into food. The greatest emphasis in safety training should be directed toward inhalation. Toxicity through lung absorption contributed to nearly 90% of toxic poisoning cases.

One of the leading causes of dermatitis comes from the use of organic solvents which may come in contact with the skin. The gastrointestinal absorption of toxic materials results in only minor concerns in manufacturing or production. Packaging of foodstuffs or pharmaceuticals must consider FDA approved polymers and additives. It is these additives and poor polymer selection which may allow toxic poisoning of consumers by ingestion.

Chlorinated hydrocarbons, trichloroethane, methyl alcohol, methylene chloride, chlorobenzene, and hydrogenated cyclic hydrocarbons are considered highly toxic. Toxicity refers to the potential for harm arising from the action of specific chemicals or other agents on living tissue. Unfortunately, there is little factual information on the hazards to humans from chemical exposures. Most data assume that humans may be as sensitive to a chemical substance as test animals. It is also apparent that not all individuals react in the same manner to the same amount of harmful material. The following toxicity rating has been used to classify toxic exposures:

Common Toxicity Rating Term	Possible Lethal Dose for Human Adults (mg/kg)
Supertoxic	< 5
Extremely toxic	5–50
Very toxic	50–500
Moderately toxic	500–5,000
Slightly toxic	5,000–15,000
Practically nontoxic	>15,000

It should be cautioned that not all individuals react to harmful materials in the same manner.

The toxicity and health hazards of selected materials are shown in Table 7-1.

It would be impossible to name, let alone discuss, the toxicological properties of all chemical compounds used by the polymer industry. The EPA has a list of toxic chemicals. Toxic poisoning is based on the rate of intake, onset of symptoms, and duration of symptoms. Acute toxicity is typically characterized by rapid absorption and exposure to the substance. Chronic toxicity implies absorption of a harmful material over a long period of time. Narcosis may produce a gradual paralysis of the central nervous system. Some of the effects may not be permanent but may be addictive. Discomfort refers to short exposure times with little exposure to danger. Many may cause skin irritation or dermatitis.

Many of the chemical hazards to personnel can be overcome by proper storage, labeling of components, and requiring that personal protective ware (gloves, respirators) be worn. Isolated storage of hazardous materials, architectural designs for safety, adequate ventilation, and a professional safety routine of accident prevention are essential.

Codes and regulations may not address all the safety and health aspects of chemical storage or toxicology, particularly as new chemicals come into more common use.

Although many particles are only a nuisance, some may cause accumulated damage to the lungs. Exposure to prolonged levels of asbestos, cotton, and silica dust may result in asbestosis, pneumoconiosis, and silicosis, respectively. Workers grinding, mixing, or cutting fibrous-reinforced or filled polymers may run a health risk if not properly protected. There has been some evidence to suggest that asbestos is carcinogenic. Composites with metallic plies or fillers of lead, zinc, or magnesium may expose workers to irritant fumes

Table 7-1. Toxicity and Health Hazards of Selected Materials

Material	Toxicity	Flash Point (°C)	Health Hazard
Acetone (dimethyl ketone)	Narcosis	−18	Skin irritation, moderately toxic inhalation
Acrylonitrile (vinyl cyanide)	Chronic	5	Absorbed through skin
Amyl acetate (banana oil)	Discomfort	25	Irritation to nose and throat from inhalation
Asbestos	Chronic	—	Respiratory diseases, carcinogenic
Benzene (benzol)	Chronic	11	Poisoning by inhalation, skin irritation and burns
Bisphenol A	Discomfort	—	Skin and nasal irritation
Boron fibers	Discomfort	—	Irritant, respiratory discomfort
Carbon dioxide	Chronic	Gas	Asphyxiation possible, chronic poisoning in small amounts
Carbon fibers	Discomfort	—	Irritant, respiratory discomfort
Carbon monoxide	Acute	Gas	Asphyxiation
Carbon tetrachloride	Acute	—	Inhaled and absorbed, chronic poisoning in small amounts
Chlorine	Acute	—	Bronchial distress, poisoning and chronic effects
Chlorobenzene (phenyl chloride)	Narcosis	29	Absorbed and inhaled, paralyzing in acute poisoning
Cobalt	Acute	—	Possible pneumoconiosis and dermatitis
Cyclohexane	Chronic	−20	Liver and kidney damage
Cyclohexanol	Discomfort	66	Inhaled and absorbed, possible organ damage
Cyclohexanone	Discomfort	44	Inhaled and absorbed, possible organ damage
Diatomaceous earth	Chronic	—	Produces silicosis

Continued

Table 7-1. Continued

Material	Toxicity	Flash Point (°C)	Health Hazard
Dichlorobenzene	Chronic	66	Possible liver damage
1,2-Dichloroethane (ethylene dichloride)	Narcosis	13	Anesthetic and narcotic, possible nerve damage
Epichlorohydrin	Acute	95	Highly irritating to eyes and respiratory tract
Ethanol (methyl alcohol)	Acute	13	Possible liver damage, narcotic effects
Fluorine	Chronic	—	Respiratory distress, acute in high concentration
Formaldehyde	Acute	—	Skin and bronchial irritations
Glass fibers	Discomfort	—	Irritant, respiratory discomfort
Isopropyl alcohol	Discomfort	14	Narcotic, irritation to respiratory tract, dermatitis
Methanol (methyl alcohol)	Chronic	11	Chronic if inhaled, poisoning may cause blindness
Methyl acrylates	Acute	3	Inhaled and absorbed, liver, kidney, and intestinal damage
Methylene chloride (dichloromethane)	Narcosis	Gas	Possible liver damage, narcotic, severe skin irritation, moderately toxic, inhalation, ingestion
Methyl ethyl ketone	Chronic	−6	Slightly toxic, effects disappearing after 48 h
Mica	Chronic	—	Pneumoconiosis, respiratory discomfort
Nickel	Acute	—	Chronic eczema, possibly carcinogenic
Phenol (carbolic acid)	Acute	80	Inhaled, absorbed through skin, narcotic, tissue damage, skin irritation
Phosgene	Acute	—	Lung damage
Pyridine	Acute	20	Liver and kidney damage

Table 7-1. Continued

Material	Toxicity	Flash Point (°C)	Health Hazard
Silane	Acute	—	Inhalation, organ damage
Silica	Chronic	—	Respiratory discomfort, silicosis, potential carcinogen
Toluene	Acute	4	Similar to benzene, possible liver damage
Vinyl acetate	Discomfort	8	No health hazard noted

and dust. Prolonged inhalation and heavy exposure can result in lead poisoning and metal fume fever.

Occupational health experts agree that the obvious and best way to prevent respiratory problems is to cut down on the amount of airborne particulates in the work environment. (See Figure 7-1.)

The hazards of fire are associated with fumes, smoke, and gases produced from combustion, thermal decomposition, and the possible resulting explosion. Flammable liquids (gasoline, acetone, toluene) are more easily ignited than combustible ones (naptha, mineral spirits, fuel oil). Flammable liquids have flash points below 100°F,

Figure 7-1. Note the personal protective ware used by these workers as they route a composite wing. (Grumman Aerospace)

while combustible liquids have flash points above 100°F. Dust explosions can occur from the ignition of fine dust particles from machining composites. Any chemical with a low flash point must be handled with care. Equipment, motors, and other electrical equipment may require explosion proof covers or properly grounded static discharge plates to prevent the possibility of vapor or dust ignition. Over 40% of all fires in industry are attributed to electrical (23%) or smoking (18%) sources.

Natural and synthetic organic materials such as wood, paper, polymers, wool, and cotton burn and produce toxic fumes. The principal killer in most fires is carbon monoxide. Heat (direct burns), oxygen deficiency, smoke, and panic are other contributing factors that can result in death or incapacitation. Care must be taken in the extrapolation of toxic fire data because laboratory tests cannot duplicate all combustion or biological factors. The actual composite product should be tested to estimate the toxic potential of burning.

Some hazards are produced from the manufacture, processing, and fabrication of composites. In the process of mixing chemicals, additives, catalysts, and so on, workers are exposed to potentially toxic substances. Many of these solvents, diluents, plasticizers, and others present fire hazards. During processing and polymerization some polymers emit toxic fumes. (See Table 7-2.)

Physical hazards exist whenever equipment is operated. The very nature of the operation may expose workers to ionizing, ultraviolet, microwave, or thermal radiation.

All machining operations are potential physical hazards including shearing, sawing, laser, and water-jet cutting.

Biological or biomechanical hazards may be associated with poor working conditions. The combined effects of poor ventilation, low lighting, repetitive motions, noise, or lack of ergonomic considerations can result in accidents or permanent biological injury. Shielding and monitoring noise and radiation sources are necessary. One of the most common occupational health claims is from various types of dermatitis. Heat, cold, chemical agents, dusts, biological agents (infections), and mechanical abrasions are common causes.

Protective barrier systems are used to protect workers from minor skin irritants and to curb the incidence of dermatitis. Personal protective ware and machine guards must be used to protect workers. Fixed barrier guards, dual hand controls, pullback devices, electronic sensing devices, and other items help to protect workers from point-of-operation hazards.

Table 7-2. Health Hazards Associated with Processing and Fabrication

Material	Health Hazard
Acrylics	Avoid skin contact with monomers and catalysts; may cause dermatitis. Normal processing temperatures cause no problems.
Acrylonitrile	Avoid skin contact with monomers. Causes dermatitis and respiratory irritation. No particular problems at normal processing temperatures.
Alkyds	Dermatitis may develop.
Aminos	Urea– and melamine–formaldehyde resins cause dermatitis and respiratory discomfort. Dust may produce dermatitis and discomfort.
Cellulose acetate	No particular problems at processing temperatures. Toxic when decomposed.
Cellulose acetate butyrate	No particular problems at processing temperatures. Toxic when decomposed.
Epoxies	Dermatitis from resin and hardener. Fumes very toxic.
Fluorocarbons	When overheated, cause polymer fume fever with flulike symptoms.
Phenolics	Phenol may cause burns and may be absorbed through the skin. Formaldehyde is irritating. Phenolics cause dermatitis and respiratory hazards.
Polyamides	Contact dermatitis has been reported but is rare.
Polyester	Monomers, promoters, and catalysts cause irritation and respiratory problems. Dermatitis may be severe. Fumes very toxic.
Polyethylene	No problems at processing temperatures.
Polystyrene	Monomers are dangerous, but plastics cause no particular problems at processing temperatures. Styrene monomers and methyl chloride very toxic.
Polyurethanes	Isocyanates are irritants to skin and respiratory tract. Avoid contact with catalysts and blowing agents. Phosgene gas, toluene vapors, and isocyanate extremely toxic.
Silicones	No problems reported. Avoid contact with hardeners with RTV rubbers.
Vinyls	Greatest danger from additives. Chlorine gas may cause discomfort.

SOURCES OF SAFETY INFORMATION

The following alphabetical sources of service organizations, standards and specifications groups, trade associations, professional societies and U.S. governmental agencies may be contacted for professional safety information, data, or further assistance:

American Chemical Society
1155 16th Street, NW
Washington, DC 20036
(202) 872-4600

American Conference of Governmental Industrial Hygienists
   (ACGIH)
6500 Glenway Avenue
Cincinnati, OH 45201
(513) 661-7881

American Industrial Hygiene Association (AIHA)
66 S. Miller Road
Akron, OH 44130
(216) 762-7294

American Insurance Association (AIA)
85 John Street
New York, NY 10038
(212) 669-0400

American Medical Association (AMA)
535 N. Dearborn Street
Chicago, IL 60610
(312) 645-5003

American National Standards Institute
1430 Broadway
New York, NY 10018
(212) 354-3300

American Society for Testing and Materials (ASTM)
1916 Race Street
Philadelphia, PA 19103
(215) 299-5400

(The) American Society of Mechanical Engineers (ASME)
United Engineering Center
345 E. 47th Street

New York, NY 10017
(212) 705-7722

American Society of Safety Engineers
850 Busse Highway
Park Ridge, IL 60068
(312) 692-4121

Chemical Manufacturers Association
2501 M Street, NW
Washington, DC 20037
(202) 887-1100

Department of Transportation
Hazardous Materials Transportation
400 7th Street, SW
Washington, DC 20590
(202) 426-4000

Environmental Protection Agency
401 M Street, SW
Washington, DC 20460
(202) 829-3535

Factory Mutual Engineering Corporation
1151 Providence Highway
Norwood, MA 02062
(617) 762-4300

Federal Emergency Management Agency
P.O. Box 8181
Washington, DC 20024
(202) 646-2500

Food and Drug Administration
200 Independence Avenue
Washington, DC 20204
(202) 245-6296

Industrial Health Foundation Inc. (IHF)
34 Penn Circle
Pittsburgh, PA 15232
(412) 363-6600

Manufacturing Chemists Association, Inc.
1825 Connecticut Avenue, NW
Washington, DC 20009
(202) 887-1100

National Association of Manufacturers
1776 F Street, NW
Washington, DC 20006
(202) 737-8551

National Fire Protection Association
470 Atlantic Avenue
Boston, MA 02210
(617) 770-3000

National Institute for Occupational Safety and Health (NIOSH)
U.S. Department of Health, Education, and Welfare
Parklawn Building
5600 Fishers Lane
Rockville, MD 20852
(301) 472-7134

National Safety Council
444 N. Michigan Avenue
Chicago, IL 60611
(312) 527-4800

Occupational Safety and Health Administration (OSHA)
U.S. Department of Labor
Department of Labor Building
Connecticut Avenue, NW
Washington, DC 20210
(202) 523-9361

Safety Standards
U.S. Department of Labor
Government Printing Office (GPO)
Washington, DC 20402
(202) 783-3238

(The) Society of the Plastics Industry, Inc. (SPI)
1025 Connecticut Avenue, NW
Suite 409
Washington, DC 20036
(202) 822-6700

Underwriters Laboratories (UL)
333 Pfingston Road
Northbrook, IL 60062
(312) 272-8800

Appendix A:
Acronyms and Abbreviations

ABS	Acrylonitrile–butadiene–styrene
ACS	Acrylonitrile–chlorinated polyethylene–styrene
AES	Acrylonitrile–ethylpropylene–styrene
AI	Amide–imide polymers
AMMA	Acrylonitrile–methyl–methacrylate
AN	Acrylonitrile
ASA	Acrylic–styrene–acrylonitrile
AU	Polyester polyurethane (copolymer elastomer)
BFK	Boron fiber-reinforced plastic
BMC	Bulk molding compounds
CA	Cellulose acetate
CAB	Cellulose acetate–butyrate
CAP	Cellulose acetate propionate
CAR	Carbon fiber
CF	Cresol–formaldehyde
CFRP	Carbon fiber-reinforced plastics
CMC	Carboxymethyl cellulose
CN	Cellulose nitrate
CP	Cellulose propionate
CPE	Chlorinated polyethylene
CPET	Crystallized PET
CPVC	Chlorinated polyvinyl chloride

CS	Casein
CTFE	(Poly) Chlorotrifluoroethylene
DAIP	Diallyl isophthalate resin
DAP	Diallyl phthalate resin
DCPD	(Poly) Dicyclopentadiene
DMC	Dough molding compound
DP	Degree of polymerization
EC	Ethyl cellulose
ECTFE	Ethylene–chlorotrifluoroethylene
EEA	Ethylene–ethyl acrylate
EMA	Ethylene–methyl acrylate
EP	Epoxy
EPE	Epoxy resin ester
EP–G–G	Prepreg of epoxy resin and glass fabric
EPM	Ethylene propylene (copolymer elastomer)
EPR	Ethylene and propylene (copolymer elastomer)
EPS	Expanded polystyrene
ETFE	EThylene–tetrafluoroethylene
EU	Polyether polyurethane
EVA	Ethylene–vinyl acetate
FEP	Fluorinated ethylene propylene
FRP	Glass fiber-reinforced polyester
FRTP	Fiber glass-reinforced thermoplastics
GF	Glass fiber reinforced
GF–EP	Glass fiber-reinforced epoxy resin
GR	Glass fiber reinforced
GRP	Glass-reinforced plastics
HIPS	High-impact polystyrene
HMW–HDPE	High molecular weight–high density polyethylene
IPN	Interpenetrating polymer network
IR	Polyisoprene
LCP	Liquid crystal polymers
LIM	Liquid impingement molding
LLDPE	Linear low-density polyethylene
LRM	Liquid reaction molding
MF	Melamine–formaldehyde

NR	Natural rubber
OPP	Oriented polypropylene
OPVC	Oriented polyvinylchloride
OSA	(Olefin-modified) Styrene–acrylonitrile
PA	Polyamide
PAA	Polyacrylic acid
PAI	Polyamide–imide
PAN	Polyacrylonitrile
PAPI	Polymethylene polyphenyl isocyanate
PB	Polybutylene
PBAN	Polybutadiene–acrylonitrile
PBS	Polybutadiene–styrene
PBT	Polybutylene terephthalate
PC	Polycarbonate
PCTFE	Polymonochlorotrifluoroethylene
PDAP	Polydiallyl phthalate
PE	Polyethylene
PEEK	Polyetheretherketone
PES	Polyether sulfone
PET	Polyethylene terephthalate
PETP	Polyethylene terephthalate
PF	Phenol–formaldehyde resin
PFA	Perfluoroalkoxy
PMCA	Polymethylchloroacrylate
PMMA	Polymethyl methacrylate
POM	Polyoxymethylene
PP	Polypropylene
PPC	Polyphthalate carbonate
PPE	Polyphenylene ether
PPO	Polyphenylene oxide
PPS	Polyphenylene sulfide
PS	Polystyrene
PSO	Polysulfone
PTFE	Polytetrafluoroethylene
PTMT	Polytetramethylene terephthalate
PU	Polyurethane
PUR	Polyurethane rubber (elastomer)

PVAc	Polyvinyl acetate
PVAl	Polyvinyl alcohol
PVB	Polyvinyl butyral
PVC	Polyvinyl chloride
PVDC	Polyvinylidene chloride
PVDF	Polyvinylidene fluoride
PVF	Polyvinyl fluoride
SAN	Styrene–acrylonitrile
SBP	Styrene–butadiene plastics
SI	Silicone
SMA	Styrene–maleic anhydride
SMC	Sheet molding compounds
SRP	Styrene–rubber plastics
TPE	Thermoplastic elastomer
TPX	Polymethylpentene
UF	Urea–formaldehyde
UHMWPE	Ultrahigh molecular weight polyethylene
UP	Urethane plastics
VCP	Vinyl chloride propylene
VDC	Vinylidene chloride
VLDPE	Very-low-density polyethylene

Appendix B:
Temperature Conversion Chart

Temperature Conversion Chart[a]

°C		°F	°C		°F	°C		°F
− 17.8	0	32	− 1.11	30	86.0	15.6	60	140.0
− 17.2	1	33.8	− 0.56	31	87.8	16.1	61	141.8
− 16.7	2	35.6	0	32	89.6	16.7	62	143.6
− 16.1	3	37.4	0.56	33	91.4	17.2	63	145.4
− 15.6	4	39.2	1.11	34	93.2	17.8	64	147.2
− 15.0	5	41.0	1.67	35	95.0	18.3	65	149.0
− 14.4	6	42.8	2.22	36	96.8	18.9	66	150.8
− 13.9	7	44.6	2.78	37	98.6	19.4	67	152.6
− 13.3	8	46.4	3.33	38	100.4	20.0	68	154.4
− 12.8	9	48.2	3.89	39	102.2	20.6	69	156.2
− 12.2	10	50.0	4.44	40	104.0	21.1	70	158.0
− 11.7	11	51.8	5.00	41	105.8	21.7	71	159.8
− 11.1	12	53.6	5.56	42	107.6	22.2	72	161.6
− 10.6	13	55.4	6.11	43	109.4	22.8	73	163.4
− 10.0	14	57.2	6.67	44	111.2	23.3	74	165.2
− 9.44	15	59.0	7.22	45	113.0	23.9	75	167.0
− 8.89	16	60.8	7.78	46	114.8	24.4	76	168.8
− 8.33	17	62.6	8.33	47	116.6	25.0	77	170.6
− 7.78	18	64.4	8.89	48	118.4	25.6	78	172.4
− 7.22	19	66.2	9.44	49	120.2	26.1	79	174.2
− 6.67	20	68.0	10.0	50	122.0	26.7	80	176.0
− 6.11	21	69.8	10.6	51	123.8	27.2	81	177.8
− 5.56	22	71.6	11.1	52	125.6	27.8	82	179.6
− 5.00	23	73.4	11.7	53	127.4	28.3	83	181.4
− 4.44	24	75.2	12.2	54	129.2	28.9	84	183.2
− 3.89	25	77.0	12.8	55	131.0	29.4	85	185.0
− 3.33	26	78.8	13.3	56	132.8	30.0	86	186.8
− 2.78	27	80.6	13.9	57	134.6	30.6	87	188.6
− 2.22	28	82.4	14.4	58	136.4	31.1	88	190.4
− 1.67	29	84.2	15.0	59	138.2	31.7	89	192.2

°C		°F	°C		°F	°C		°F
32.2	90	194.0	160	320	608	349	660	1220
32.8	91	195.8	166	330	626	354	670	1238
33.3	92	196.7	171	340	644	360	680	1256
33.9	93	199.4	177	350	662	366	690	1274
34.4	94	201.2	182	360	680	371	700	1292
35.0	95	203.0	188	370	698	377	710	1310
35.6	96	204.8	193	380	716	382	720	1328
36.1	97	206.6	199	390	734	388	730	1346
36.7	98	208.4	204	400	752	393	740	1364
37.2	99	210.2	210	410	770	399	750	1382
37.8	100	212.0	216	420	788	404	760	1400
38	100	212	221	430	806	410	770	1418
43	110	230	227	440	824	416	780	1436
49	120	248	232	450	842	421	790	1454
54	130	266	238	460	860	427	800	1472
60	140	284	243	470	878	432	810	1490
66	150	302	249	480	896	438	820	1508
71	160	320	254	490	914	443	830	1526
77	170	338	260	500	932	449	840	1544
82	180	356	266	510	950	454	850	1562
88	190	374	271	520	968	460	860	1580
93	200	392	277	530	986	466	870	1598
99	210	410	282	540	1004	471	880	1616
100	212	413	288	550	1022	477	890	1634
104	220	428	293	560	1040	482	900	1652
110	230	446	299	570	1058	488	910	1670
116	240	464	304	580	1076	493	920	1688
121	250	482	310	590	1094	499	930	1706
127	260	500	316	600	1112	504	940	1724
132	270	518	321	610	1130	510	950	1742
138	280	536	327	620	1148	516	960	1760
143	290	554	332	630	1166	521	970	1778
149	300	572	338	640	1184	527	980	1796
154	310	590	343	650	1202	532	990	1814

[a]Celsius (°C) = $\frac{5}{9}$ (°F − 32); Fahrenheit (°F) = $\frac{9}{5} \times$ °C + 32

To convert a given temperature, either fahrenheit or Celsius, find its value in center column and its Celsius or fahrenheit conversion in the appropriate column. For example, 90°F is 32.2°C and 90°C is 194°F.

Appendix C: Weight, Length, and Volume Equivalences

Weight of 1000 Pieces in Pounds
Based on Weight of One Piece in Grams

Weight per Piece in Grams	Weight per 1000 Pieces in Pounds	Weight per Piece in Grams	Weight per 1000 Pieces in Pounds	Weight per Piece in Grams	Weight per 1000 Pieces in Pounds
1	2.2	27	59.4	53	116.7
2	4.4	28	61.6	54	118.9
3	6.6	29	63.8	55	121.1
4	8.8	30	66.0	56	123.3
5	11.0	31	68.2	57	125.5
6	13.2	32	70.4	58	127.7
7	15.4	33	72.6	59	129.9
8	17.6	34	74.8	60	132.1
9	19.8	35	77.0	61	134.3
10	22.0	36	79.2	62	136.5
11	24.2	37	81.4	63	138.7
12	26.4	38	83.7	64	140.9
13	28.6	39	85.9	65	143.1
14	30.8	40	88.1	66	145.3
15	33.0	41	90.3	67	147.5
16	35.2	42	92.5	68	149.7
17	37.4	43	94.7	69	151.9
18	39.6	44	96.9	70	154.1
19	41.8	45	99.1	71	156.3
20	44.0	46	101.3	72	158.5
21	46.2	47	103.5	73	160.7
22	48.4	48	105.7	74	162.9
23	50.6	49	107.9	75	165.1
24	52.8	50	110.1	76	167.4
25	55.0	51	112.3	77	169.6
26	57.2	52	114.5	78	171.8

Weight per Piece in Grams	Weight per 1000 Pieces in Pounds	Weight per Piece in Grams	Weight per 1000 Pieces in Pounds	Weight per Piece in Grams	Weight per 1000 Pieces in Pounds
79	174.0	87	191.6	95	209.2
80	176.2	88	193.8	96	211.4
81	178.4	89	196.0	97	213.6
82	180.6	90	198.2	98	215.8
83	182.8	91	200.4	99	218.0
84	185.0	92	202.6	100	220.2
85	187.2	93	204.8	. . .	. . .
86	189.4	94	207.0	. . .	. . .

Equivalent weights: 1 g = 0.0353 oz.; 0.0625 lb = 1 oz = 28.3 g; 454 g = 1 lb

Length Equivalents (Millimeters to Inches)

Millimeters	Inches	Millimeters	Inches	Millimeters	Inches
1	0.03937	34	1.33860	67	2.63779
2	0.07874	35	1.37795	68	2.67716
3	0.11811	36	1.41732	69	2.71653
4	0.15748	37	1.45669	70	2.75590
5	0.19685	38	1.49606	71	2.79527
6	0.23622	39	1.53543	72	2.83464
7	0.27559	40	1.57480	73	2.87401
8	0.31496	41	1.61417	74	2.91338
9	0.35433	42	1.65354	75	2.95275
10	0.39370	43	1.69291	76	2.99212
11	0.43307	44	1.73228	77	3.03149
12	0.47244	45	1.77165	78	3.07086
13	0.51181	46	1.81102	79	3.11023
14	0.55118	47	1.85039	80	3.14960
15	0.59055	48	1.88976	81	3.18897
16	0.62992	49	1.92913	82	3.22834
17	0.66929	50	1.96850	83	3.26771
18	0.70866	51	2.00787	84	3.30708
19	0.74803	52	2.04724	85	3.34645
20	0.78740	53	2.08661	86	3.38582
21	0.82677	54	2.12598	87	3.42519
22	0.86614	55	2.16535	88	3.46456
23	0.90551	56	2.20472	89	3.50393
24	0.94488	57	2.24409	90	3.54330
25	0.98425	58	2.28346	91	3.58267
26	1.02362	59	2.32283	92	3.62204
27	1.06299	60	2.36220	93	3.66141
28	1.10236	61	2.40157	94	3.70078
29	1.14173	62	2.44094	95	3.74015
30	1.18110	63	2.48031	96	3.77952
31	1.22047	64	2.51968	97	3.81889
32	1.25984	65	2.55905	98	3.85826
33	1.29921	66	2.59842	99	3.89763
				100	3.93700

Volume equivalents: 1 cm^3 = 0.061 in.^{3}; 1 in.^{3} = 16.387 cm^3

Appendix D:
Inch Equivalents and Draft Angles

Decimal Equivalents of Fractions of One Inch					
1/64	0.015625	23/64	0.359375	11/16	0.687500
1/32	0.031250	3/8	0.375000	45/64	0.703125
3/64	0.046875	25/64	0.390625	23/32	0.718750
1/16	0.062500	13/32	0.406250	47/64	0.734375
5/64	0.078125	27/64	0.421875	3/4	0.750000
3/32	0.093750	7/16	0.437500	49/64	0.765625
7/64	0.109375	29/64	0.453125	25/32	0.781250
1/8	0.125000	15/32	0.468750	51/64	0.796875
9/64	0.140625	31/64	0.484375	13/16	0.812500
5/32	0.156250	1/2	0.500000	53/64	0.828125
11/64	0.171875	33/64	0.515625	27/32	0.843750
3/16	0.187500	17/32	0.531250	55/64	0.859375
13/64	0.203125	35/64	0.546875	7/8	0.875000
7/32	0.218750	9/16	0.562500	57/64	0.890625
15/64	0.234375	37/64	0.578125	29/32	0.906250
1/4	0.250000	19/32	0.593750	59/64	0.890625
17/64	0.265625	39/64	0.609375	15/16	0.937500
9/32	0.281250	5/8	0.625000	61/64	0.953125
19/64	0.296875	41/64	0.640625	31/32	0.968750
5/16	0.312500	21/32	0.656250	63/64	0.984375
21/64	0.328125	43/64	0.671875	1	1.000000
11/32	0.343750				

Standard Draft Angles

Depth	1/4°	1/2°	1°	1 1/2°	2°	2 1/2°
1/32	0.0001	0.0003	0.0005	0.0008	0.0011	0.0014
1/16	0.0003	0.0006	0.0011	0.0016	0.0022	0.0027
3/32	0.0004	0.0008	0.0016	0.0025	0.0033	0.0041
1/8	0.0005	0.0010	0.0022	0.0033	0.0044	0.0055
3/16	0.0008	0.0016	0.0033	0.0049	0.0065	0.0082
1/4	0.0011	0.0022	0.0044	0.0066	0.0087	0.0109
5/16	0.0014	0.0027	0.0055	0.0082	0.0109	0.0137
3/8	0.0016	0.0033	0.0065	0.0098	0.0131	0.0164
7/16	0.0019	0.0038	0.0076	0.0115	0.0153	0.0191
1/2	0.0022	0.0044	0.0087	0.0131	0.0175	0.0218
5/8	0.0027	0.0054	0.0109	0.0164	0.0218	0.0273
3/4	0.0033	0.0065	0.0131	0.0196	0.0262	0.0328
7/8	0.0038	0.0076	0.0153	0.0229	0.0306	0.0382
1	0.0044	0.0087	0.0175	0.0262	0.0349	0.0437
1 1/4	0.0055	0.0109	0.0218	0.0327	0.0437	0.0546
1 1/2	0.0064	0.0131	0.0262	0.0393	0.0524	0.0655
1 3/4	0.0076	0.0153	0.0305	0.0458	0.0611	0.0764
2	0.0087	0.0175	0.0349	0.0524	0.0698	0.0873

3°	5°	7°	8°	10°	12°	15°	Depth
0.0016	0.0027	0.0038	0.0044	0.0055	0.0066	0.0084	1/32
0.0033	0.0055	0.0077	0.0088	0.0110	0.0133	0.0168	1/16
0.0049	0.0082	0.0115	0.0132	0.0165	0.0199	0.0251	3/32
0.0066	0.0109	0.0153	0.0176	0.0220	0.0266	0.0335	1/8
0.0098	0.0164	0.0230	0.0263	0.0331	0.0399	0.0502	3/16
0.0131	0.0219	0.0307	0.0351	0.0441	0.0531	0.0670	1/4
0.0164	0.0273	0.0384	0.0439	0.0551	0.0664	0.0837	5/16
0.0197	0.0328	0.0460	0.0527	0.0661	0.0797	0.1005	3/8
0.0229	0.0383	0.0537	0.0615	0.0771	0.0930	0.1172	7/16
0.0262	0.0438	0.0614	0.0703	0.0882	0.1063	0.1340	1/2
0.0328	0.0547	0.0767	0.0878	0.1102	0.1329	0.1675	5/8
0.0393	0.0656	0.0921	0.1054	0.1322	0.1595	0.2010	3/4
0.0459	0.0766	0.1074	0.1230	0.1543	0.1860	0.2345	7/8
0.0524	0.0875	0.1228	0.1405	0.1763	0.2126	0.2680	1
0.0655	0.1094	0.1535	0.1756	0.2204	0.2657	0.3349	1 1/4
0.0786	0.1312	0.1842	0.2108	0.2645	0.3188	0.4019	1 1/2
0.0917	0.1531	0.2149	0.2460	0.3085	0.3720	0.4689	1 3/4
0.1048	0.1750	0.2456	0.2810	0.3527	0.4251	0.5359	2

Appendix E: Specific Gravity

Specific Gravities of Materials

Material	Specific Gravity
Aluminum	2.7
Asbestos	2.0–2.8
Brass	8.5
Brick	1.4–2.2
Celluloid	1.4
Cellulose acetate	1.27–1.63
Cold-molded compound	2.0
Copper	8.8
Rogers DAP/DAIP	1.36–1.80
Glass	2.4–5.9
Hard rubber	1.1–1.4
Hard wood	0.65–1.23
India rubber	0.91
Iron	7.0–7.9
Ivory	1.83–1.92
Lead	11.3
Marble	2.6–2.8
Methyl methacrylate	1.18
Phenolics	1.25–2.08
Melamine–formaldehyde	1.47–1.52
Polystyrene	1.05–1.07
Porcelain	2.5
Shellac	1.6
Soft wood	0.38–0.92
Steel	7.6–7.8
Water	1.0
Zinc	7.1

Conversion of Specific Gravity to Grams per Cubic Inch
($16.39 \times$ Specific Gravity $=$ Grams/In.^{3})

Specific Gravity	g/in.^{3}	Specific Gravity	g/in.^{3}
1.20	19.7	1.82	29.8
1.22	20.0	1.84	30.2
1.24	20.3	1.86	30.5
1.26	20.7	1.88	30.8
1.28	21.0	1.90	31.1
1.30	21.3	1.92	31.5
1.32	21.6	1.94	31.8
1.34	22.0	1.96	32.1
1.36	22.3	1.98	32.5
1.38	22.6	2.00	32.8
1.40	22.9	2.02	33.1
1.42	23.3	2.04	33.4
1.44	23.6	2.06	33.8
1.46	23.9	2.08	34.1
1.48	24.3	2.10	34.4
1.50	24.6	2.12	34.7
1.52	24.9	2.14	35.1
1.54	25.2	2.16	35.4
1.56	25.6	2.18	35.7
1.58	25.9	2.20	36.1
1.60	26.2	2.22	36.4
1.62	26.6	2.24	36.7
1.64	26.9	2.26	37.0
1.66	27.2	2.28	37.4
1.68	27.5	2.30	37.7
1.70	27.9	2.32	38.0
1.72	28.2	2.34	38.4
1.74	28.5	2.36	38.7
1.76	38.8	2.38	39.0
1.78	29.2	2.40	39.3
1.80	29.5		

To determine the cost/in.^{3}: Price/lb $\times$ specific gravity $\times$ 0.03163 $=$ \$1.32 $\times$ 1.76 $\times$ 0.03163 $=$ \$0.09/in.^{3}

Appendix F:
Circles and Temperatures/Pressures

Diameters and Areas of Circles

Diam. (in.)	Area	Diam. (in.)	Area	Diam. (in.)	Area
1/64	0.00019	27/32	0.55914	5/16	4.2000
1/32	0.00077	7/8	0.60132	3/8	4.4301
3/64	0.00173	29/32	0.64504	7/16	4.6664
1/16	0.00307	15/16	0.69029	1/2	4.9087
3/32	0.00690	31/32	0.73708	9/16	5.1572
1/8	0.01227			5/8	5.4119
5/32	0.01917	1-	0.7854	11/16	5.6727
3/16	0.02761	1/16	0.8866	3/4	5.9396
7/32	0.03758	1/8	0.9940	13/16	6.2126
1/4	0.04909	3/16	1.1075	7/8	6.4918
9/32	0.06213	1/4	1.2272	15/16	6.7771
5/16	0.07670	5/16	1.3530		
11/32	0.09281	3/8	1.4849	3-	7.0686
3/8	0.11045	7/16	1.6230	1/16	7.3662
13/32	0.12962	1/2	1.7671	1/8	7.6699
7/16	0.15033	9/16	1.9175	3/16	7.9798
15/32	0.17257	5/8	2.0739	1/4	8.2958
1/2	0.19635	11/16	2.2465	5/16	8.6179
17/32	0.22165	3/4	2.4053	3/8	8.9462
9/16	0.24850	13/16	2.5802	7/16	9.2806
19/32	0.27688	7/8	2.7612	1/2	9.6211
5/8	0.30680	15/16	2.9483	9/16	9.9678
21/32	0.33824			5/8	10.321
11/16	0.37122	2-	3.1416	11/16	10.680
23/32	0.40574	1/16	3.3410	3/4	11.045
3/4	0.44179	1/8	3.5466	13/16	11.416
25/32	0.47937	3/16	3.7583	7/8	11.793
13/16	0.51849	1/4	3.9761	15/16	12.177

Diam. (in.)	Area	Diam. (in.)	Area	Diam. (in.)	Area
4-	12.566	1/4	30.680	1/4	82.516
1/16	12.962	3/8	31.919	3/8	84.541
1/8	13.364	1/2	33.183	1/2	86.590
3/16	13.772	5/8	34.472	5/8	88.664
1/4	14.186	3/4	35.785	3/4	90.763
5/16	14.607	7/8	37.122	7/8	92.886
3/8	15.033				
7/16	15.466	7-	38.485	11-	95.033
1/2	15.904	1/8	39.871	1/2	103.87
9/16	16.349	1/4	41.282		
5/8	16.800	3/8	42.718	12-	113.10
11/16	17.257	1/2	44.179	1/2	122.72
3/4	17.721	5/8	45.664		
13/16	18.190	3/4	47.173	13-	132.73
7/8	18.665	7/8	48.707	1/2	143.14
15/16	19.147				
		8-	50.265	14-	153.94
5-	19.635	1/8	51.849	1/2	165.13
1/16	20.129	1/4	53.456		
1/8	20.629	3/8	55.088	15-	176.71
3/16	21.125	1/2	56.745	1/2	188.69
1/4	21.648	5/8	58.426		
5/16	22.166	3/4	60.132	16-	201.06
3/8	22.691	7/8	61.862	1/2	213.82
7/16	23.211				
1/2	23.758	9-	63.617	17-	226.98
9/16	24.301	1/8	65.397	1/2	240.53
5/8	24.850	1/4	67.201		
11/16	25.406	3/8	69.029	18-	254.47
3/4	25.967	1/2	70.882	1/2	268.80
13/16	26.535	5/8	72.760		
7/8	27.109	3/4	74.662	19-	283.53
15/16	27.688	7/8	76.589	1/2	298.65
6-	28.274	10-	78.540	20-	314.16
1/8	29.465	1/8	80.516	1/2	330.06

Steam Temperature Versus Gauge Pressure

Steam Temperature Versus Gauge Pressure	
Gauge Pressure (lb)	Temperature (°F)
50	297.5
55	302.4
60	307.1
65	311.5
70	315.8
75	319.8
80	323.6
85	327.4
90	331.1
95	334.3
100	337.7
105	341.0
110	344.0
115	347.0
120	350.0
125	353.0
130	356.0
135	358.0
140	361.0
145	363.0
150	365.6
155	368.0
160	370.3
165	372.7
170	374.9
175	377.2
180	379.3
185	381.4
190	383.5
195	385.7
200	387.5

Bibliography

Accident Prevention Manual for Industrial Operations. Chicago: National Safety Council, 1974.

"A Guide to Statically Conductive and EMI Attenuating Composites," Bulletin 223-1184, LNP Corporation.

Bauer, R.S., ed. *Epoxy Resin Chemistry II.* American Chemical Society Symposium, No. 221, p. 320.

Berngardt, Ernest. *CAE Computer Aided Engineering for Injection Molding.* New York: Hanser Publishers, 1983.

Billmeyer, Fred W. *Textbook of Polymer Science,* 3rd ed. New York: Wiley, 1984.

Brooke, Lindsay. "Cars of 2000: Tomorrow Rides Again!" *Automotive Industries,* pp. 50–67 (May 1986).

Broutman, L., and R. Krock. *Composite Materials,* 6 vols. New York: Academic Press.

Budinski, Kenneth. *Engineering Materials: Properties and Selection,* 2nd ed. Reston, VA: Reston Publishing Company, 1983.

Carraher, Charles E., Jr., and James Moore. *Modification of Polymers.* New York: Plenum Press, 1983.

"Chemical Emergency Preparedness Program Interim Guidance," Revision 1, #9223.01A. Washington, DC: United States Environmental Protection Agency, 1985.

Clauser, H.R. "Advanced Composite Materials," *Scientific American,* 229 (2):36–44 (July 1983).

Composite Materials Technology. Warrendale, PA: Society of Automotive Engineers, 1986.

"Defense Standardization Manual: Defense Standardization and

Specification Program Policies, Procedures and Instruction," DOD 4120.3-M, August 1978.

Delmonte, John. *Technology of Carbon and Graphite Fiber Composites.* New York: Wiley-Interscience, 1982.

Dreger, Donald. "Design Guidelines of Joining Advanced Composites," *Machine Design*, pp. 89–93 (May 8, 1980).

Dym, Joseph. *Product Design with Plastics: A Practical Manual.* New York: Industrial Press, 1983.

Ehrenstein, G., and G. Erhard. *Designing with Plastics: A Report on the State of the Art.* New York: Hanser Publishers, 1984.

Engle, E., and others. *Atlas of Polymer Damage.* Englewood Cliffs, NJ: Prentice-Hall, 1981.

English, Lawrence. "Liquid-Crystal Polymers: In a Class of Their Own," *Manufacturing Engineering*, pp. 36–41 (March 1986).

English, Lawrence. "The Expanding World of Composites," *Manufacturing Engineering*, pp. 27–31 (April 1986).

Fitts, Bruce. "Fiber Orientation of Glass Fiber-Reinforced Phenolics," *Materials Engineering*, pp. 18–22 (November 1984).

Folkes, M. *Short Fibre Reinforced Thermoplastics.* New York: Wiley, 1982.

Gaylord, H.R. "Advanced Composite Materials," *Scientific American*, 229 (2):36–44 (July 1973).

Grayson, Martin. *Encyclopedia of Composite Materials and Components.* New York: Wiley, 1984.

Hall, Christopher. *Polymer Materials: An Introduction for Technologists and Scientists.* New York: Wiley, 1981.

Hartwig, G., and D. Evans, eds. *Nonmetallic Materials and Composites at Low Temperatures (2).* New York: Plenum Press, 1982, p. 412.

Hench, Larry, and D. Ulrich. *Ultrastructure Processing of Ceramics, Glasses, and Composites.* New York: Wiley, 1984.

Holmes, M., and D.J. Just. *GRP in Structural Engineering.* City: Applied Science Publishers, 1983, p. 298.

Jayne, Benjamin. *Theory and Design of Wood and Fiber Composite Materials.* Syracuse: Syracuse University Press, 1972.

Johnson, Wayne, and R. Schwed. "Computer-Aided Design and Drafting," *Engineered Systems*, pp. 48–51 (March/April 1986).

Kliger, Howard. "Customizing Carbon-Fiber Composites: For Strong, Rigid, Lightweight Structures," *Machine Design*, pp. 150–157 (December 6, 1979).

Levy, Sidney, and J. Harry Dubois. *Plastics Product Design Engineering Handbook*, 2nd ed. New York: Chapman and Hall, 1984.

Lubin, George. *Handbook of Composites.* New York: Van Nostrand Reinhold Company, 1982.

Metcalfe, A. "Interfaces in Metal Matrix Composites," *Composite Materials.* New York: Academic Press, 1974.

Miller, Edward. *Plastics Products Design Handbook*: Part A. New York: Marcel Dekker, 1981.

Miller, Edward. *Plastics Products Design Handbook*: Part B. New York: Marcel Dekker, 1983.

Modern Plastics Encyclopedia, Vol 62 (10A), October 1985.

Mohr, G., and others. *SPI Handbook of Technology and Engineering of Reinforced Plastics/Composites*, 2nd ed. Malabar, FL: Krieger Publishing Company, 1984.

Moore, G.R., and D.E. Kline. *Properties and Processing of Polymers for Engineers.* Englewood Cliffs, NJ: Prentice-Hall, 1984.

Naik, Saurabh, and others. "Evaluating Coupling Agents for Mica/Glass Reinforcement of Engineering Thermoplastics," *Modern Plastics*, pp. 1979–1980 (June 1985).

Newby, Gregory, and John Theberge. "Long-Term Behavior of Reinforced Thermoplastics," *Machine Design*, pp. 171–177 (March 8, 1984).

"Noryl Resin Design Guide," CDX-830, Selkirk: General Electric Company, 1985.

Powell, Peter C. *Engineering with Polymers.* New York: Chapman and Hall, 1983.

Richardson, Terry. *A Guide to Metrics.* Ann Arbor, MI: Prakken Publications, 1978.

Richardson, Terry. *Industrial Plastics: Theory and Application.* Cincinnati: South-Western Publishing Company, 1983.

Rosen, Stephen. *Fundamental Principles of Polymeric Materials.* New York: Wiley, 1982.

Schwartz, M.M. *Composite Materials Handbook.* New York: McGraw-Hill, 1984.

Serferis, J.C., ed. *The Role of the Polymeric Matrix in the Processing and Structural Properties of Composite Materials.* New York: Plenum Press, 1983, p. 650.

Seymour, Ramold B., and Charles Carraher. *Polymer Chemistry.* New York: Marcel Dekker, 1981.

Shan, Vishu. *Handbook of Plastics Testing Technology.* New York: Wiley, 1984.

"Solve Barrel and Screw Wear Problems with Tungsten Carbide Composites," *Plastics Technology*, pp. 30–32 (September 1985).

Smith, O. *The Science of Engineering Materials*, 3rd ed. Englewood Cliffs, NJ: Prentice-Hall, 1986.

"Standardization Case Studies: Defense Standardization and Specification Program," Department of Defense, March 17, 1986.

Stepek, J., and H. Daoust. *Additives for Plastics*. New York: Springer-Verlag, 1983, p. 260.

Stille, John. *Introduction to Polymer Chemistry*. New York: Wiley, 1966.

Szycher, M. *Biocompatible Polymers, Metals and Composites*. Lancaster, PA: Technomic Publishing Company, 1983.

Tewary, V. *Mechanics of Fibre Composites*. New York: Wiley, 1978.

Tough Composite Materials: Recent Developments. Park Ridge, NJ: Noyes Publications, 19XX.

"Torlon Engineering Polymers/Design Manual," Bulletin TAT-35, Chicago: Amoco Chemicals Corporation, 1984.

Von Hassell, Agostino. "Computer Integrated Manufacturing: Here's How to Plan for It," *Plastics Technology Productivity Series*, No. 1, 1986.

Wigotsky, Victor. "Plastics Are Making Dream Cars Come True," *Plastics Engineering*, pp. 19–27 (May 1986).

Wigotsky, Victor. "U.S. Moldmakers Battle Foreign Prices for Survival," *Plastics Engineering*, pp. 22–23 (November 1985).

Wood, Stuart. "Patience: Key to Big Volume in Advanced Composites," *Modern Plastics*, pp. 44–48 (March 1986).

Index